REVISION NOTES IN PHYSICS

Book Two

BOOKS BY M. NELKON

Published by Heinemann

ADVANCED LEVEL PRACTICAL PHYSICS (SI) *(with J. Ogborn)*
SCHOLARSHIP PHYSICS (SI)
OPTICS, SOUND AND WAVES (SI)
MECHANICS AND PROPERTIES OF MATTER (SI)
PRINCIPLES OF ATOMIC PHYSICS AND ELECTRONICS
REVISION NOTES IN PHYSICS (SI)
Book I, Mechanics, Electricity, Atomic Physics
Book II, Optics, Waves, Sound, Heat, Properties of Matter
GRADED EXERCISES IN PHYSICS (SI)
TEST PAPERS IN PHYSICS (SI)
REVISION BOOK IN ORDINARY LEVEL PHYSICS (SI)
ELEMENTARY PHYSICS, Book I and II *(with A. F. Abbott)*
AN INTRODUCTION TO THE MATHEMATICS OF PHYSICS *(with J. H. Avery)*
ELECTRONICS AND RADIO *(with H. I. Humphreys)*
SOLUTIONS TO ADVANCED LEVEL PHYSICS QUESTIONS (SI)
SOLUTIONS TO ORDINARY LEVEL PHYSICS QUESTIONS (SI)

Published by Chatto & Windus

PRINCIPLES OF PHYSICS (O-Level, SI)
FUNDAMENTALS OF PHYSICS (O-Level, SI)
EXERCISES IN ORDINARY LEVEL PHYSICS *(with* SI *units)*
C.S.E. PHYSICS
SI UNITS: AN INTRODUCTION FOR ADVANCED LEVEL

Published by Edward Arnold

ELECTRICITY (Advanced Level, SI)

Published by Blackie

HEAT (Advanced Level, SI)

BOOKS BY P. PARKER

Published by Heinemann

INTERMEDIATE HEAT
ELECTRICITY AND MAGNETISM

Published by Arnold

ELECTRONICS

REVISION NOTES IN PHYSICS

FOR ADVANCED LEVEL AND INTERMEDIATE STUDENTS

by

M. NELKON M.Sc.(Lond.), F. Inst.P., A.K.C.

Formerly Head of the Science Department, William Ellis School, London

Book Two

OPTICS, WAVES, SOUND, HEAT, PROPERTIES OF MATTER

THIRD EDITION

(with SI units)

HEINEMANN EDUCATIONAL BOOKS
LONDON

Heinemann Educational Books Ltd
LONDON EDINBURGH MELBOURNE AUCKLAND TORONTO
SINGAPORE HONG KONG KUALA LUMPUR
IBADAN NAIROBI JOHANNESBURG
LUSAKA NEW DELHI

ISBN 0 435 68641 0

First published 1953
Reprinted 1954, 1955, 1956, 1957, 1958,
1959, 1960, 1962, 1964
Second Edition 1965
Reprinted 1966
Reprinted 1969 with amendments
Third Edition 1971 (with SI units)
Reprinted 1973, 1974

Published by Heinemann Educational Books Ltd
48 Charles Street, London W1X 8AH
Printed Offset Litho and bound in Great Britain by
Cox & Wyman Ltd, London, Fakenham and Reading

PREFACE

These Notes are intended to assist students taking the Advanced level or Intermediate examination in Physics. They are compiled from many years' experience of teaching and lecturing, and enumerate, compare, and contrast the principal points in Optics, Waves, Sound, Heat and Properties of Matter to this standard. The book is intended to supplement the existing text-books of Physics, and as it provides a summary of the work in a two-year Advanced level or Intermediate course, it is hoped that the Notes will prove useful to the second-year student for quick revision. Some degree of over-simplification has been unavoidable in condensing the subject-matter. Worked examples, taken from examination questions of recent years, are included in illustration of the subject-matter.

The author is much indebted to Mr. A. F. Abbot, formerly Senior Science Master, Latymer Upper School, London, for reading the manuscript and for valuable suggestions; to Dr. J. Duffey, formerly of Watford Technical College, and Mr. C. F. Whorwell, William Ellis School, London, for reading the proofs; and to the following Examining Boards, for permission to reprint questions: University of London (*L.*), Joint Matriculation Board (*N.*), Oxford and Cambridge Joint Board (*O. & C.*), Cambridge Local Examinations Syndicate (*C.*), Joint Welsh Board (*W.*).

THIRD EDITION

This edition covers the new Advanced level syllabus of the Examining Boards in Optics, Waves, Sound, Heat and Properties of Matter. In the revised text, emphasis has been placed on optical instruments, waves and physical optics; the kinetic theory of gases and conduction in heat; and molecular theory in properties of matter. SI units only are now used throughout the text, in accordance with the future policy of the Boards.

I am indebted to R. P. T. Hills, St. John's College, Cambridge, for his valuable assistance with this edition.

TO THE STUDENT

This book provides a quick revision of Optics, Waves, Sound, Heat and Properties of Matter. It does not replace any of your text-books on Physics, but supplements them. *Diagrams of the apparatus should be drawn when revising experiments.*

The following technique will be found useful when answering descriptive questions in Physics: (1) Draw a diagram first, and letter the various parts. (2) Then give the description or account, referring to the diagram by the letters. (3) If a measurement is required, *always give the formula for calculating the quantity concerned.* The formula should not be proved unless this is asked for in the question.

Contents

OPTICS

Optics

GEOMETRICAL OPTICS

1. REFLECTION AT PLANE AND CURVED MIRRORS

PLANE MIRRORS

Laws of reflection. (1) The incident, reflected ray, and the normal all lie in the same plane. (2) The angle of incidence = the angle of reflection.

Images. A real object has a virtual image in a plane mirror. A *virtual object* O (due to a converging incident beam) has a real image I. See Fig. 1.1

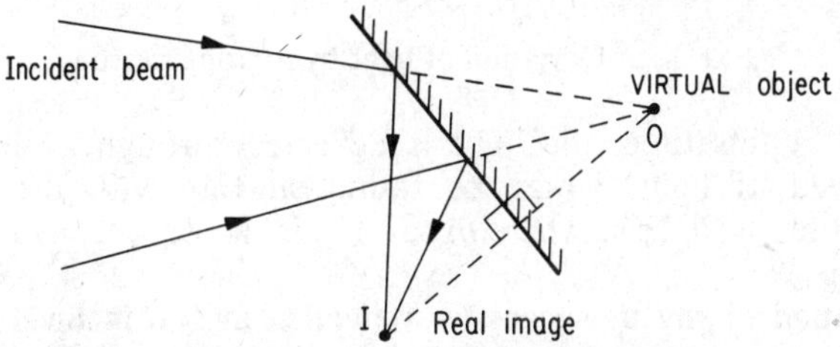

FIG. 1.1. Plane mirror—virtual object, real image.

Deviation of a reflected ray by rotating mirror. With incident ray direction constant, deviation = 2θ, where θ is the angle of rotation. (This is proved most easily by working with the *glancing angle*, the angle between the incident ray and the mirror, and utilizing the standard result that:

$$\text{Deviation of light by a mirror} = 2 \times \text{glancing angle.})$$

Mirror galvanometer. The incident beam on the mirror is deviated through *twice* the angle of rotation θ of the mirror attached to the coil. Fig. 1.2*a*. D is usually 1 metre (1000 mm) and the scale deflection d is relatively small. Hence 2θ (radians) = d/D to good approximation.

Speed of light. In the Foucault method of measuring the speed of light (p. 49), a light beam, reflected from a rotating plane mirror M, is incident on a concave mirror C. Fig. 1.2*b*. The light arrives back at M after travelling a distance $2r$. If M has rotated through an

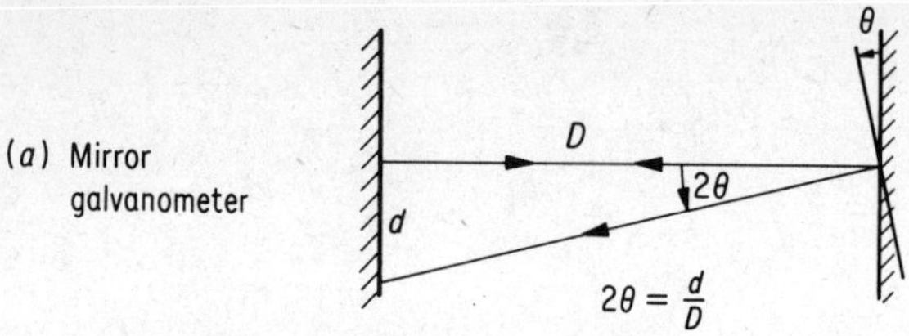

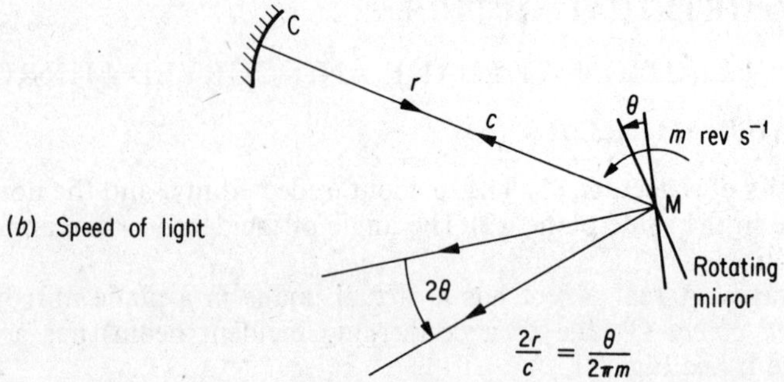

FIG. 1.2. Deviation of light by rotating mirror.

angle θ in this time, the light is deflected through an angle 2θ. The speed of light c may be found starting with the relation $2r/c = \text{time} = \theta/2\pi m$, where m rev s^{-1} is the rate of revolution of M. See p. 49.

Deviation of ray by successive reflection at two inclined mirrors.

$$\text{Deviation} = 2\theta,$$

where θ is the angle between the mirrors. This is proved by finding the deviation at one mirror (using the glancing angle formula), then at the second mirror, and adding or subtracting the two deviations.

Sextant. This has one fixed and one movable mirror. The fixed mirror is divided vertically into a clear and a silvered half. The former sights on the horizon; the latter reflects light from the distant sun or star when the movable mirror is turned. When the distant object (star or sun) is brought to horizon level viewed through the clear glass,

$$\text{angle of elevation} = 2 \times \text{angle of rotation of mirror}.$$

CURVED MIRRORS

Concave (converging) mirror. (1) Has a real focus ($f = r/2$).

(2) Produces real and inverted images when the object is further from the mirror than its focal length.

(3) At the centre of curvature C, an object O_1 produces a real inverted image I_1 of equal length at the same place. Fig. 1.3.

(4) Produces *virtual*, erect, and magnified image I_2 when the object, O_2, is *nearer* the mirror than its focal length. Fig. 1.3.

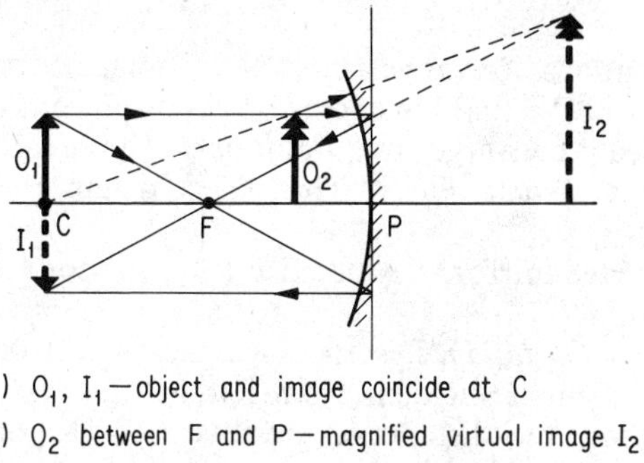

FIG. 1.3. Images in concave mirror.

Convex (diverging) mirror. (1) Has a virtual focus ($f = r/2$).

(2) Produces virtual, erect, and diminished images (driving mirror) and has a wide field of view. Fig. 1.4*a*.

(3) Note that a virtual object O (*converging* incident beam) may produce a real image I. Fig. 1.4*b*.

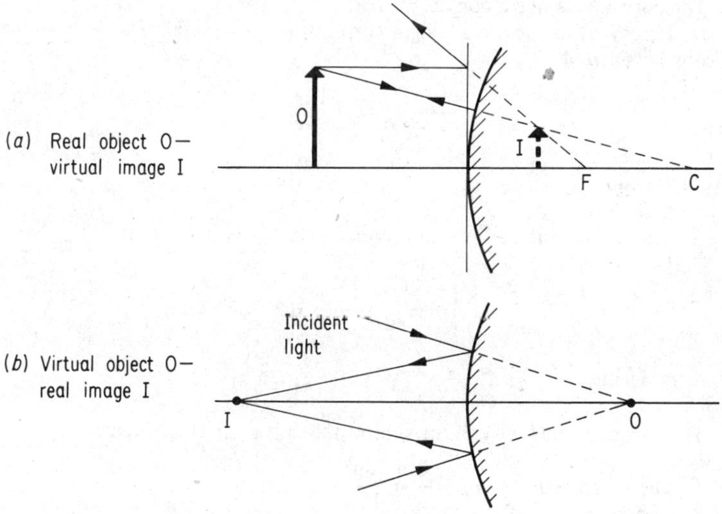

FIG. 1.4. Images in convex mirror.

Mirror formulæ.

$$\frac{1}{v}+\frac{1}{u} = \frac{1}{f}$$

$$\text{Linear (transverse) magnification, } m, = \frac{v}{u} = \frac{v}{f}-1.$$

Sign conventions. *Real is positive.* "Real" image or object distance is positive (+). "Virtual" image or object distance is negative (−).

When a beam converges to a point *behind* a mirror, that point is regarded as a *virtual object*. The object distance is then negative (−).

For a concave mirror, f is positive (+); for convex mirror, f is negative (−).

[*New Cartesian.* Distances are measured from the mirror, and the light is incident from left to right. To the right of the mirror, the distance is positive (+); to the left of the mirror, the distance is negative (−).

On this convention f is negative for a concave mirror and positive for a convex mirror.]

EXAMPLE

Two small illuminated objects are placed 15 cm apart in a line perpendicular to the visual axis of, and 60 cm in front of, an eye. Calculate the position and the distance between the images reflected from the surface of the cornea, which has a radius of curvature of 8 mm.

The cornea acts like a convex mirror.

In the "real is positive" sign convention, $u = +60$, $f = \frac{1}{2}r = -0{\cdot}4$. From $1/v+1/u = 1/f$,

$$\frac{1}{v}+\frac{1}{+60} = \frac{1}{-0{\cdot}4},$$

from which $$v = -\frac{60}{151}\text{ cm.}$$

[*New Cartesian* convention, $u = -60, f = +0{\cdot}4$. From $1/v+1/u = 1/f$,

$$\frac{1}{v}+\frac{1}{-60} = \frac{1}{+0{\cdot}4},$$

from which $$v = +\frac{60}{151}\text{ cm.}]$$

The image is thus 60/151 cm behind the cornea (virtual image).

$$\text{The magnification, } m, = \frac{v}{u} = \frac{60}{151} \div 60 = \frac{1}{151}$$

$$\therefore \text{ distance between images} = \frac{1}{151}\times 15 = 0{\cdot}1\text{ cm.}$$

Methods for f and r of curved mirrors

Concave mirror. (1) Coincidence of image and object with pin, or by means of a light box, provides r (Fig. 1.5a,b).

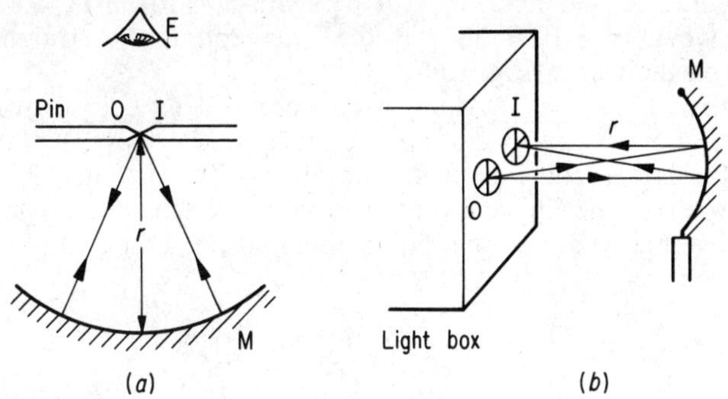

FIG. 1.5. Radius of curvature of concave mirror.

(2) Measurement of u and v; then plot $1/v$ v. $1/u$ graph, and observe the intercept on either axis, OA or OB, which gives $1/f$ (Fig. 1.6a).

To locate virtual images, (i) place an object pin nearer to the mirror than its focal length, (ii) either use a tall locator pin behind

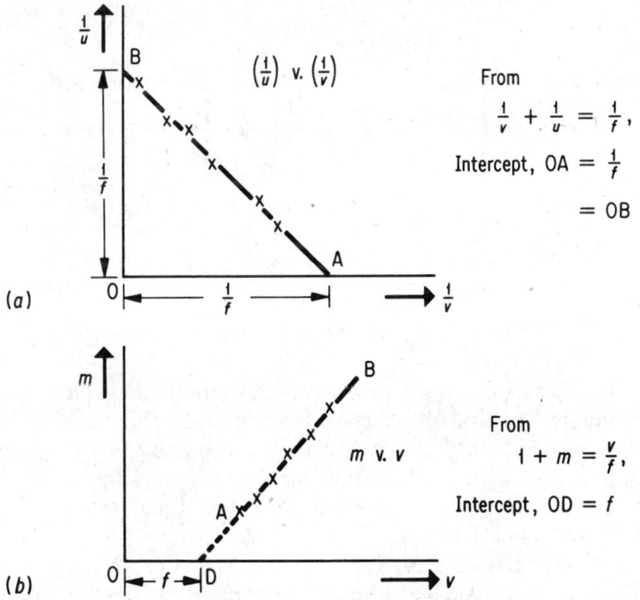

FIG. 1.6. (a) Location of virtual images (b) Magnification method.

the mirror and arrange for no parallax with the image, or use a plane mirror method with the locator pin in front of it.

(3) Magnification method. Measure linear magnification, m, for several image distances, v. Plot m against v. From $m = v/f - 1$, $v = f$ when $m = 0$. Thus f is the intercept of the straight line graph on the v-axis (Fig. 1.6*b*).

Convex mirror. (1) Arrange a converging lens L to provide an incident convergent beam on the mirror M. Move the mirror until the image coincides with the object. The distance PC from the mirror to the image I, obtained with the lens alone when the mirror is removed, is r, the radius of curvature (Fig. 1.7*a*).

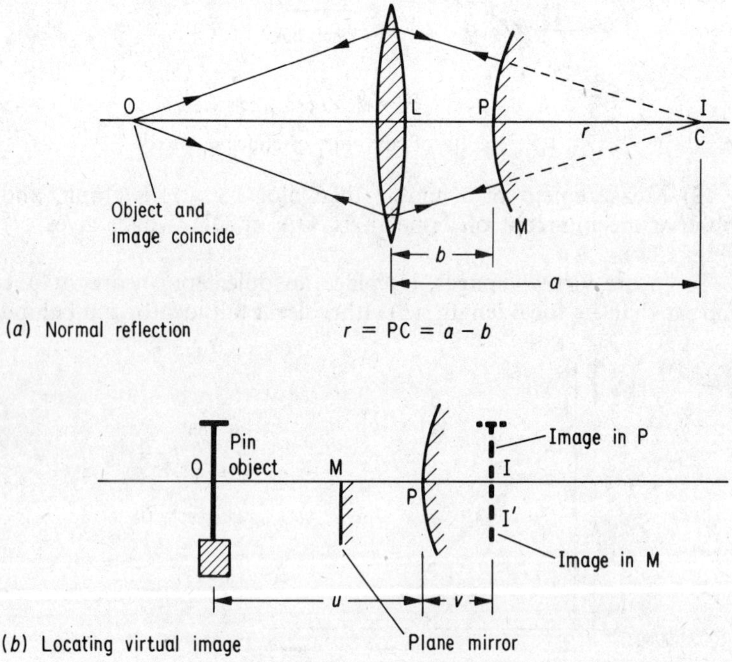

FIG. 1.7. Focal length of convex mirror.

(2) Arrange a tall object in front of the mirror P, which gives a virtual image I behind the mirror. Place a plane mirror M in front of P and move it until the image I′ in the plane mirror coincides by no parallax with that in the convex mirror (Fig. 1.7*b*). The image distance, v, can then be found. The focal length is then found from $1/v + 1/u = 1/f$.

Proof of mirror formulæ. Take a point object O on the principal axis. Draw a ray inclined at a small angle α and incident at a point P on the mirror. Join the centre of curvature C to P (this is the normal at P), and draw the reflected ray on the other side of the normal. This passes through the image, I, on the axis.

If CP, IP make angles β, γ respectively with the axis, then

$$(i)\ \alpha+\gamma = 2\beta \text{ for a concave mirror,}$$
$$(ii)\ \gamma-\alpha = 2\beta \text{ for a convex mirror,}$$

Substituting tangents for the angles, *and using the sign convention*, the formula can be derived. It should be noted that the formula holds only for rays near to the principal axis, when a unique focus is obtained. If the beam is wide, aberration occurs.

To prove the formula 'lateral magnification, m, $= v/u$', use similar triangles with the lengths of object and image in them.

General. (1) In calculations on unknown distances, let x be the *numerical* value of the unknown distance. Then the object distance $u = +x$ or $-x$ on substituting in the formula.

(2) The formula $m = \frac{v}{f} - 1$ is useful in questions on magnification.

2. REFRACTION AT PLANE SURFACES. CRITICAL ANGLE. PRISMS

REFRACTION AT PLANE SURFACES

Laws of refraction. (1) The incident, refracted ray and the normal all lie in the same plane.

(2) For two given media, $\frac{\sin i}{\sin r}$ is a constant, where i is the angle of incidence and r is the angle of refraction.

Refractive index. When light travels from a medium 1 into a medium 2, the refractive index ${}_1n_2$ can be *defined by*:

$${}_1n_2 = \frac{\sin i}{\sin r} \text{ (Fig. 2.1}a\text{)}$$

or by:

$${}_1n_2 = \frac{\text{velocity of light in medium 1}}{\text{velocity of light in medium 2}}.$$

The *absolute refractive index*, *n*, of a medium is the refractive index when the incident light is travelling in a vacuum, or air for practical purposes, and the incident light is yellow (sodium light). For crown glass $n = 1{\cdot}51$ (approx.), for flint glass $n = 1{\cdot}64$ (approx.), for water $n = 1{\cdot}33$ (approx.).

Relations between refractive indices. We denote the refractive index from water to air by ${}_wn_a$, and similarly for other refractive index values. ${}_gn_a$ is the refractive index from glass to air (Fig. 2.1*b*).

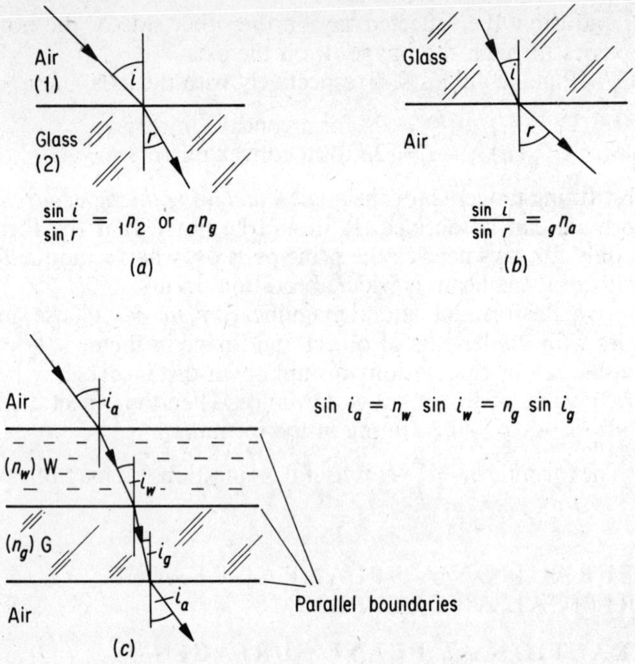

FIG. 2.1. Refraction and refractive index.

Note that

$$ {}_w n_a = \frac{1}{{}_a n_w} = \frac{1}{4/3} = \frac{3}{4}, $$

$$ {}_w n_g = {}_w n_a \times {}_a n_g = \frac{{}_a n_g}{{}_a n_w} = \frac{3/2}{4/3} = \frac{9}{8} $$

or

$$ {}_1 n_3 = {}_1 n_2 \times {}_2 n_3. $$

Light travelling from one medium to another. "$n \sin i$" is a constant quantity, where n is the absolute refractive index of that medium and i is the angle made by the ray in that medium with the normal. Thus from glass to water,

$$ n_g \sin i_g = n_w \sin i_w. $$

When light travels from air to water to glass, the boundaries of the media being parallel, "$n \sin i$" is again a constant quantity. Thus, if n_a (=1) is the refractive index of air, then, as in Fig. 2.1*c*,

$$ n_a \sin i_a = n_w \sin i_w = n_g \sin i_g. $$

Displacement and apparent depth. Apparent depth = t/n, where t is the true depth or thickness of the medium or parallel-sided block.

$$\therefore \text{ apparent displacement, } d, = t - \frac{t}{n} = t\left(1 - \frac{1}{n}\right).$$

The apparent displacement formula is independent of the position of the object when placed below a parallel-sided block of glass and viewed directly from above. This can be shown by taking rays from the object and drawing the image position.

Total internal reflection. Critical angle. Two conditions are necessary for total internal reflection to occur:

(1) The light must travel from a dense to a less dense medium.

(2) The angle of incidence in the dense medium must be greater than the critical angle, *c*, between the two media. See Fig. 2.2*a*.

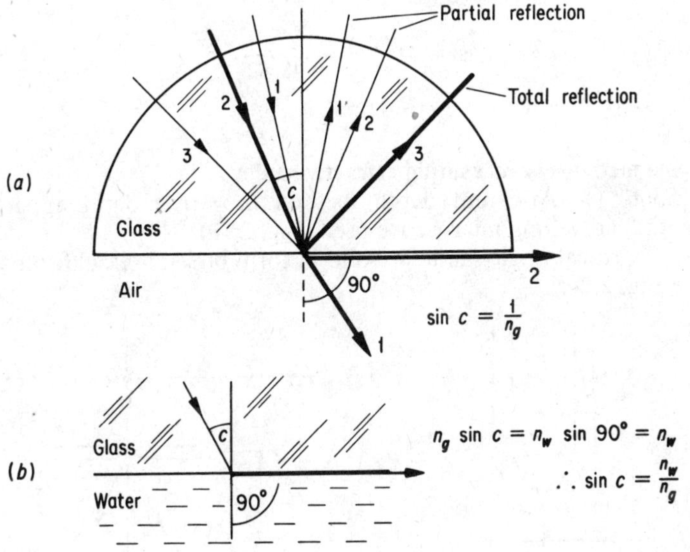

FIG. 2.2. Total internal reflection. Critical angle.

The *critical angle, c,* is defined as the angle of incidence in the dense medium when the angle of refraction in the less dense medium is 90°. Thus for air-glass

$$\sin c = \frac{1}{n_g}.$$

From glass to water, the critical angle is given by $n_g \sin c = n_w \sin 90°$ (Fig. 2.2*b*). Thus

$$\sin c = \frac{n_w}{n_g}.$$

EXAMPLE

If the critical angle is 48° 42′ when the media concerned are air and water and 36° 47′ when they are air and glass, what is it when they are water and glass?

The critical angle, c, is obtained when (*i*) light travels from glass (the denser medium) to water, and (*ii*) the angle of refraction in the water is 90°. Then, since "$n \sin i$" is a constant,

$$n_g \sin c = n_w \sin 90° = n_w \times 1$$

$$\therefore \sin c = \frac{n_w}{n_g}.$$

Now $$n_w = 1/\sin c_w = 1/\sin 48° 42',$$

and $$n_g = 1/\sin c_g = 1/\sin 36° 47'.$$

$$\therefore \sin c = \frac{n_w}{n_g} = \frac{\sin 36° 47'}{\sin 48° 42'}$$

$$\therefore c = 52° 51'.$$

Some methods of measuring refractive index

Solid. (1) Apparent depth method (n = true depth/apparent depth). Travelling microscope used (Fig. 2.3*a*).

(2) Critical angle method (solid in form of semi-circular block). See Fig. 2.2*a*.

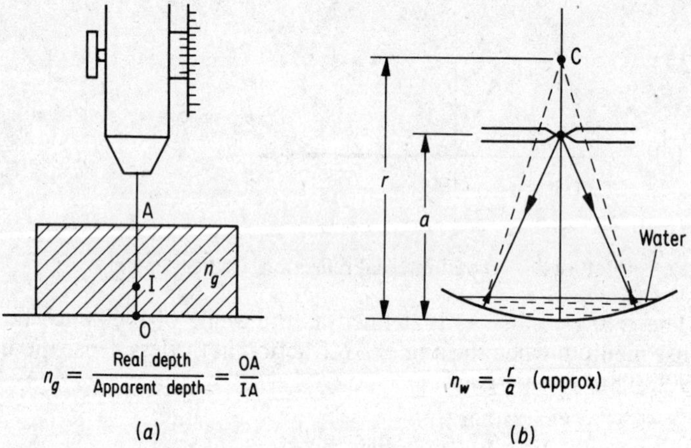

FIG. 2.3. Refractive index of (*a*) glass (*b*) small amount of liquid.

Liquid. (1) Air-cell (critical angle) method. An air film between two microscope slides X is placed in a vessel V with plane sides containing the liquid (Fig. 2.4*a*). The film is illuminated by yellow light S, which disappears at one stage as X is turned. Then n_l =

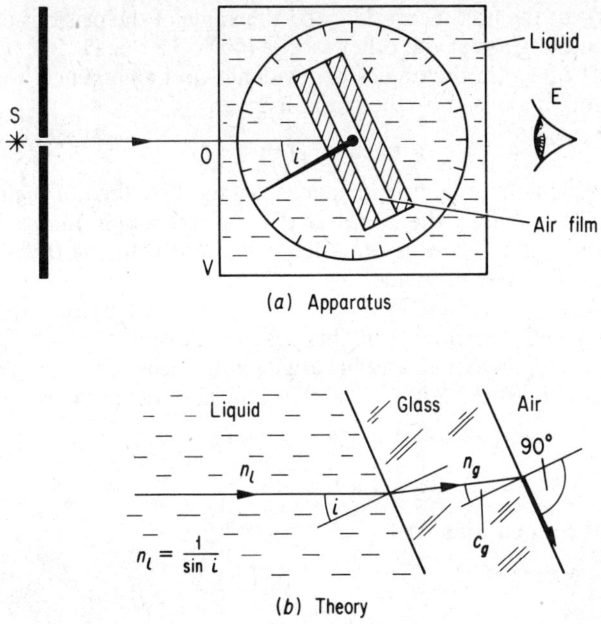

FIG. 2.4. Air-cell method for liquid.

1/sin *i*, where *i* is the angle of rotation of X from normal incidence.

Proof: When the light first disappears, the angle of refraction in the *air* of the cell is 90°. From $n \sin i$ = constant for parallel media, then $n_l \sin i = n_g \sin c_g = 1 \times \sin 90° = 1$. Thus $n_l = 1/\sin i$.

(2) Apparent depth method.

(3) Concave mirror method. Useful for *small* quantity of liquid (Fig. 2.3*b*). The radius of curvature, *r*, is first found with a pin; some liquid is then smeared on the mirror, and the new distance, *a*, from the pin to the mirror is found when coincidence is again obtained. Then $n_w = r/a$ to a good approximation.

(4) Pulfrich refractometer. Accurate method for small quantity of liquid, which is placed in a tube on a glass cube.

(5) Lens and plane mirror method. (See p. 24.)

REFRACTION THROUGH PRISMS

General. (1) When a ray is refracted through a glass prism, and r_1, r_2 are the respective angles in the glass which the ray makes with the normals at the points of incidence and emergence, then always

$$r_1 + r_2 = A,$$

where *A* is the angle of the prism. Thus if the angle of refraction in the glass at the first face is 32°, and the angle of the prism is 60°, the angle in the glass at the other face is (60° − 32°) or 28°.

(2) If i_1, i_2 are the angles of incidence and emergence, the total deviation, d, caused by the prism is given by

$$d = (i_1 - r_1) + (i_2 - r_2).$$

(3) When the angle of incidence on the first face of a prism is gradually increased, the deviation decreases to a minimum and then increases again. There is only one minimum value, corresponding to a certain angle of incidence.

Minimum deviation. This occurs when the ray passes *symmetrically* through the prism. In this case the angles of incidence and emergence, i, are equal, and the angles with the normals in the glass, r, are equal (Fig. 2.5*a*). If D is the minimum deviation value,

$$D = 2i - 2r.$$

Also,

$$r + r = A.$$

From these relations,

$$n = \frac{\sin i}{\sin r} = \frac{\sin\left(\frac{A+D}{2}\right)}{\sin\left(\frac{A}{2}\right)}.$$

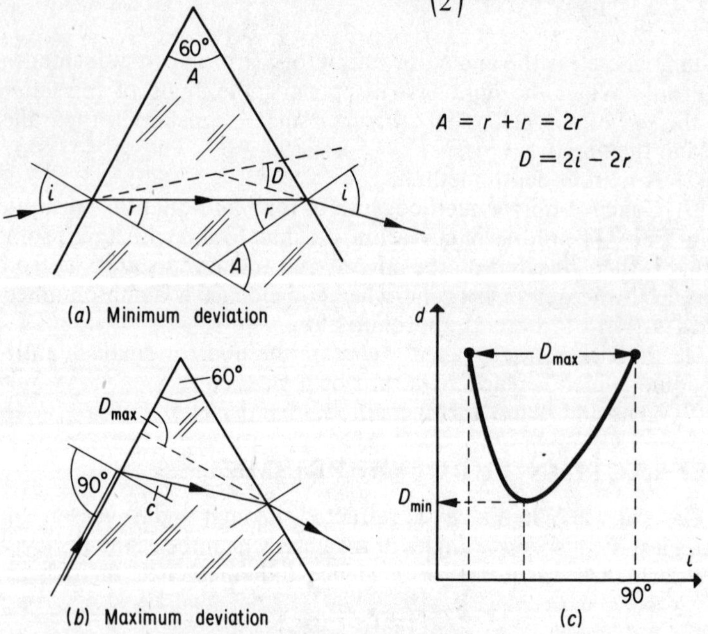

FIG. 2.5. Refraction through prism.

Maximum deviation. The maximum deviation, D_{max}, produced by a prism corresponds to light *at grazing incidence* on the prism, that is, the angle of incidence is 90° (Fig. 2.5*b*). The angle of refraction at the first face is then the critical angle value, since $\sin 90° = n \sin r$, or $\sin r = 1/n$. Fig. 2.5*c* shows the general variation of the deviation d with angle of incidence i.

EXAMPLE

The prism of a spectrometer has a refracting angle of 60° and is made of glass whose refractive indices for red and violet are respectively 1·514 and 1·530. A white source is used and the instrument is set to give minimum deviation for red light. Determine (*a*) the angle of incidence of the light on the prism, (*b*) the angle of emergence of the violet light, (*c*) the angular width of the spectrum.

First part. The optical system is the collimator, telescope, prism. Note that the emerging red rays from the prism are all parallel, as are the emerging blue rays; but the red rays are deviated less than the blue rays, and are therefore brought to a different focus.

Second part. (*a*) The minimum deviation, D, for red light is given by

$$n = \frac{\sin\left(\dfrac{A+D}{2}\right)}{\sin A/2}$$

$$\therefore \sin\left(\frac{A+D}{2}\right) = n \sin\frac{A}{2} = 1{\cdot}514 \times \sin 30° = 0{\cdot}757$$

$$\therefore \frac{A+D}{2} = 49°\,12'.$$

$$\text{But angle of incidence, } i, = \frac{A+D}{2}$$

$$\therefore i = 49°\,12'.$$

(*b*) The angle of incidence of the violet light is also 49° 12′. Hence the angle of refraction, r, in the glass is given by

$$\frac{\sin 49°\,12'}{\sin r} = n = 1{\cdot}530$$

$$\therefore \sin r = \frac{\sin 49°\,12'}{1{\cdot}530}$$

$$\therefore r = 29°\,39'$$

$$\therefore \text{angle of incidence on other face of prism} = A - r = 60° - 29°\,39' = 30°\,21'.$$

$\therefore$ angle of emergence, r_1, is given by

$$\frac{\sin r_1}{\sin 30°\,21'} = n = 1{\cdot}530$$

$$\therefore \sin r_1 = 1{\cdot}530 \times \sin 30° 21'$$

$$\therefore r_1 = 50° 38'.$$

(c) ∴ angular width of spectrum $= 50° 38' - 49° 12' = 1° 26'$.

Largest (limiting) angle of prism for refraction. In this case a ray at grazing incidence (90° to the normal) emerges along the surface of the second face (grazing emergence). Then, if c is the critical angle and A is the corresponding (limiting) angle of the prism,

$$A = c + c = 2c.$$

Total reflecting prisms. For crown glass of $n = 1{\cdot}51$, the critical angle is about 42°. An isosceles right-angle prism of crown glass thus acts as a total reflector of light if the angle of incidence is 45° in the glass.

These prisms are used in prism binoculars and in submarine periscopes.

The spectrometer. This instrument is used, among other purposes, for measuring the refractive index of glass in the form of a prism and for the examination of the colours emitted by a source of light. It consists of:

(1) A *collimator*. This provides a beam of parallel light from a narrow slit at one end.

(2) A *telescope*. This receives parallel light, and brings them to a focus on the cross-wires in the eyepiece.

(3) A *table*. A prism is placed on the table, which can be rotated about an axis perpendicular to that of the telescope (Fig. 2.6).

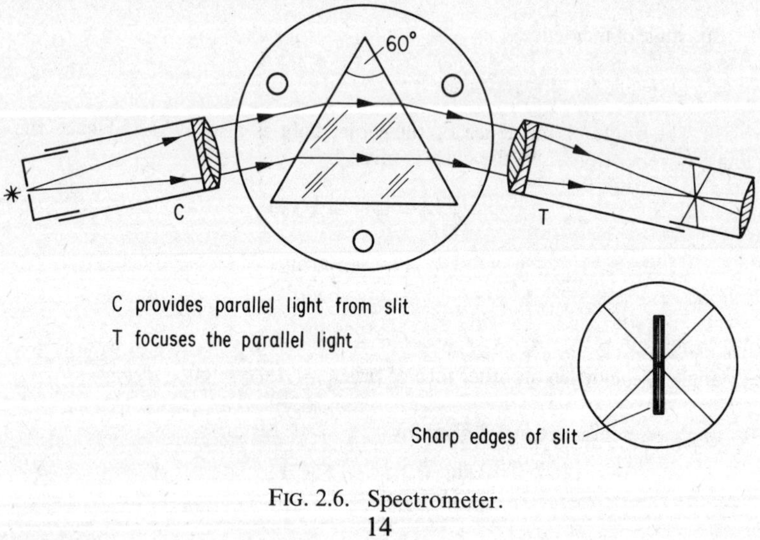

FIG. 2.6. Spectrometer.

Adjustments to spectrometer. (1) Adjust the eyepiece so that the cross-wires are clearly seen (Fig. 2.7*a*).

(2) Then adjust the telescope for parallel light by sighting on a distant object, or by sighting through a lens on a wire at its focus (Fig. 2.7*b*).

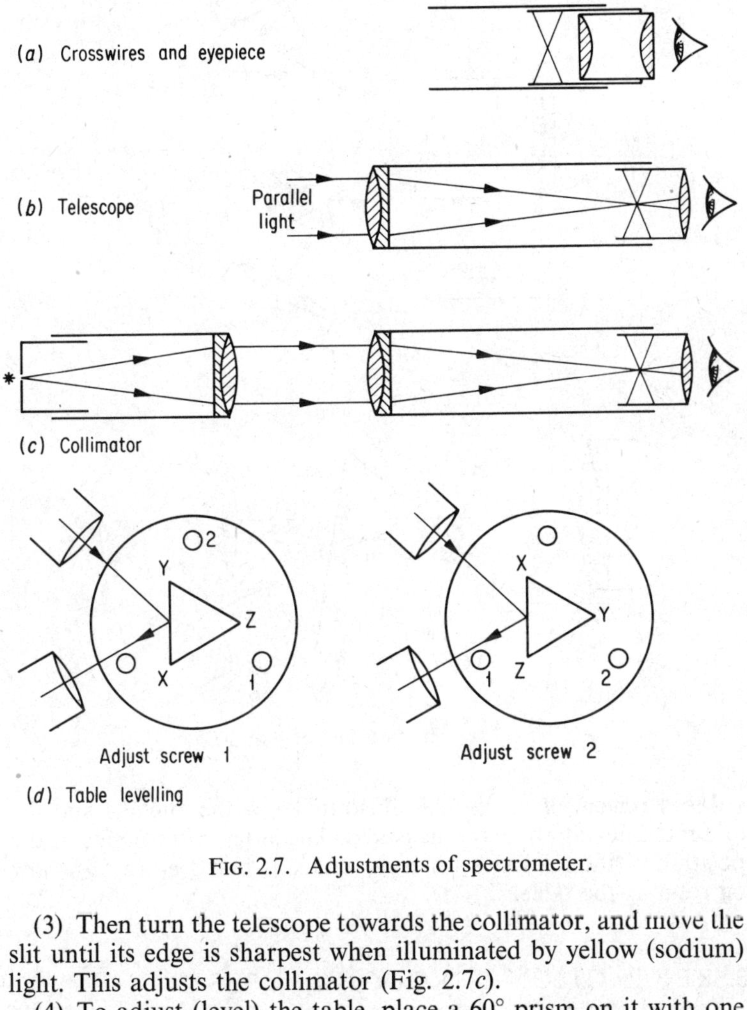

FIG. 2.7. Adjustments of spectrometer.

(3) Then turn the telescope towards the collimator, and move the slit until its edge is sharpest when illuminated by yellow (sodium) light. This adjusts the collimator (Fig. 2.7*c*).

(4) To adjust (level) the table, place a 60° prism on it with one face XY perpendicular to two of the levelling screws (Fig. 2.7*d*). Observe the slit by *reflection* at XY and adjust screw 1 until the slit is in the middle of the field of view. Keeping the telescope fixed, rotate the table until the slit is seen by reflection at the second face XZ of the prism. Now adjust the third screw, 2 say, until the image of

the slit is again in the centre of the field of view. Note that only two screws are adjusted.

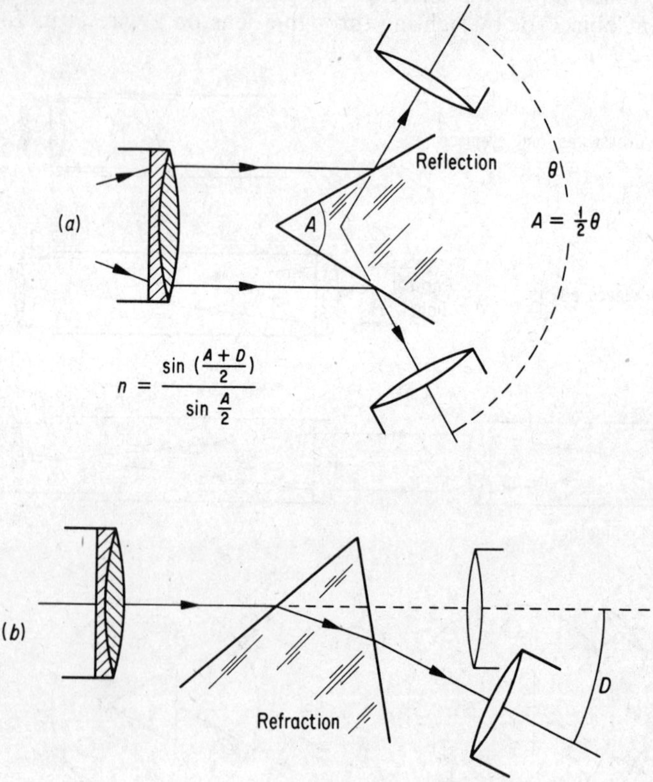

FIG. 2.8. Measurement of *A* and *D*.

Measurement of *n*. Fig. 2.8 illustrates how the angle *A* and the minimum deviation *D* are measured. The latter corresponds to the position of the telescope when the image is first seen to *turn back* on rotating the table.

3. REFRACTION THROUGH LENSES

Converging (convex) lens. (1) Has a real focus.

(2) Produces real and inverted images when the object O_1 is further from the lens than the focus F. When the object O_2 is nearer to the lens than the focal length, an erect, virtual, and magnified image is obtained (magnifying glass) (Fig. 3.1*a*).

Diverging (concave) lens. (1) Has a virtual focus.

(2) Produces an erect, virtual and diminished image of an object.

(3) A virtual object O may produce a real image, as shown in Fig. 3.1*b*.

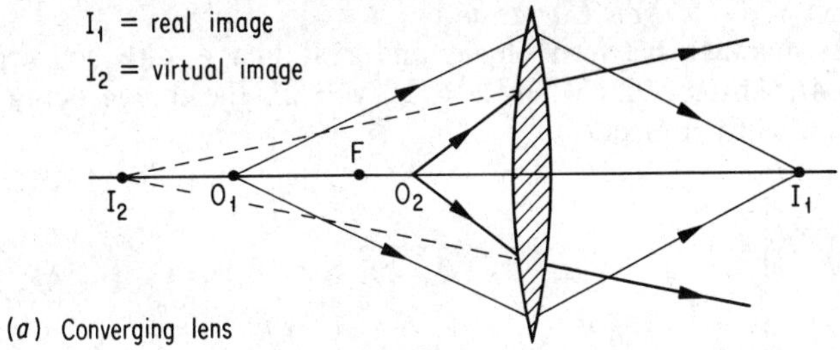

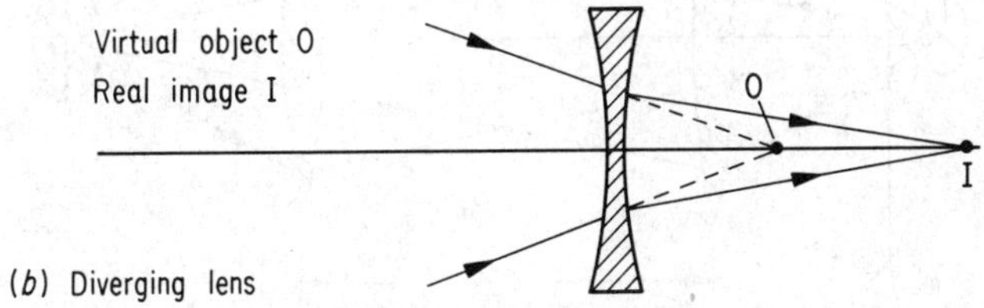

FIG. 3.1. Images in (*a*) converging (*b*) diverging lenses.

In accurate drawing of images with both types of lenses, (*a*) the lens is represented by a straight line, (*b*) a ray through the centre of the thin lens passes straight through the lens, (*c*) a ray parallel to the principal axis is refracted through the focus.

Principal foci. A lens has two principal foci; a beam parallel to the principal axis incident on the left side of the lens is refracted through one focus (called the "second principal focus"), whilst a similar beam incident on the right side of the lens is refracted through the other focus (called the "first principal focus").

Conjugate foci. If I is the image of a point object O in a lens, the points O, I are called "conjugate foci".

Lens formulæ. (1) With the "real is positive" sign convention,

$$\frac{1}{v}+\frac{1}{u}=\frac{1}{f}.$$

[*New Cartesian* sign convention, $1/v-1/u = 1/f$]

(2) For a converging (convex) lens, f is $+$.

For a diverging (concave) lens, f is $-$.

(3) The linear magnification, $m, = \dfrac{v}{u}$.

(4) $m = \frac{v}{f} - 1$ (real is positive)

[$m = 1 - v/f$, (*New Cartesian*)].

(5) $xx' = f^2$, where x, x' are the distances of object and image from their respective foci (Newton's relation).

[$xx' = -f^2$, *New Cartesian.*]

Least distance between object and real image with converging lens = $4f$. This occurs when $u = 2f$, $v = 2f$, the image being the same size as the object.

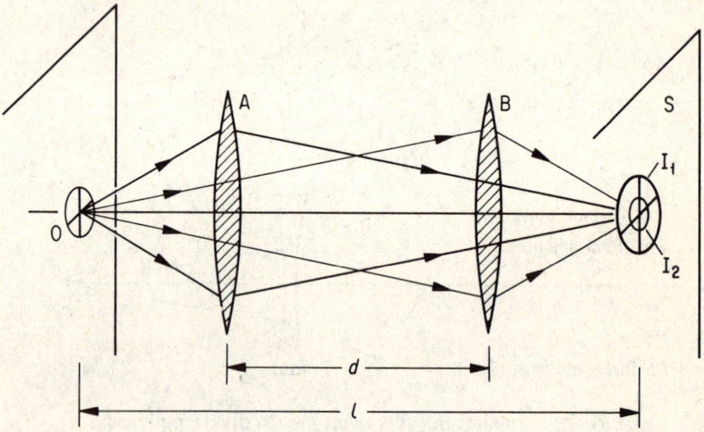

Object O and screen S fixed — two images focused on S by displacing lens

$$f = \frac{l^2 - d^2}{4l}$$

FIG. 3.2. Displacement of lens.

Object and screen fixed, and converging lens. In general, two images, I_1, I_2, are obtained on the fixed screen when the lens is moved. See Fig. 3.2. In one case, when v is greater than u, the image I_1 is larger than the object O; in the other case v is smaller than u, and the image I_2 is reduced. *The object distance in the first case = the image distance in the second case.* Thus

$$m = 1/m_1,$$

where m, m_1 are the respective linear magnifications.

If d is the displacement of the lens in this case, and l is the distance between the object and fixed screen, then $u = (l-d)/2$, $v = d+(l-d)/2 = (l+d)/2$, by considering the interchangeability of "u" and "v" in the two positions of the lens (Fig. 3.2). Substituting in the lens equation for u and v, and simplifying,

$$\therefore f = \frac{l^2 - d^2}{4l}.$$

This formula is used in finding the focal length of a lens fixed in a tube (p. 21).

Convergent beam incident on lens. When a *convergent* beam is incident on a lens, we have the case, on the "real is positive" convention, of a *virtual object*. The "object" is the point to which the beam would converge if the lens were not there, and its distance from the lens, u, is negative.

With the New Cartesian convention, the incident light being on the left of the lens, the virtual object is on the right of the lens. Thus u is positive in this case.

EXAMPLE

A thin bi-convex lens rests on a plane mirror. It is found that a point object placed 20 cm above the lens coincides with its own image. Determine the position and the nature of the image when the object is placed (*a*) 8 cm, (*b*) 12 cm from the lens-mirror combination.

The focal length, f, = 20 cm, since object and image coincide.

(*a*) With the "real is positive" convention, the image distance in the lens is given by

$$\frac{1}{v} + \frac{1}{(+8)} = \frac{1}{(+20)}.$$

$$\left[\textit{New Cartesian}\text{ convention, } \frac{1}{v} - \frac{1}{(-8)} = \frac{1}{(+20)}\right].$$

$$\therefore v = -13\tfrac{1}{3}\text{ cm.}$$

Thus a diverging beam is incident on the mirror. The mirror reflects the beam, and the reflected beam appears to come from a point $13\frac{1}{3}$ cm behind the mirror, which is the real "object" for the lens. Thus the final image position in the lens is given by

$$\frac{1}{v} - \frac{1}{(+13\frac{1}{3})} = \frac{1}{(+20)}.$$

$$\left[\textit{New Cartesian}\text{ convention, } \frac{1}{v} - \frac{1}{(-13\frac{1}{3})} = \frac{1}{(+20)}\right].$$

Solving, $\therefore v = -40$ cm.

Thus a virtual image is obtained, 40 cm from the lens.

(*b*) In this case, $\frac{1}{v} + \frac{1}{(+12)} = \frac{1}{(+20)}$.

$$\left[\textit{New Cartesian}\text{ convention, } \frac{1}{v} - \frac{1}{(-12)} = \frac{1}{(+20)}\right].$$

Solving, $\therefore v = -30$ cm.

The final image position in the lens is thus given by

$$\frac{1}{v}+\frac{1}{(+30)} = \frac{1}{(+20)}.$$

$$\left[\textit{New Cartesian convention,}\ \frac{1}{v}-\frac{1}{(-30)} = \frac{1}{(+20)}\right].$$

Solving, $\therefore v = +60$ cm.

Thus a real image is obtained in this case, 60 cm from the lens.

It is left as an exercise to the student to show that the final image is erect in (*a*), and inverted in (*b*).

Two thin lenses in contact. If f_1, f_2 are the respective focal lengths of the lenses, the combined focal length F is given by

$$\frac{1}{F} = \frac{1}{f_1}+\frac{1}{f_2}.$$

This formula is proved by taking a point object in front of the two lenses A, B, then finding the image distance v' by refraction at lens A. For refraction at lens B, v' is the (virtual) object distance. When the lens formula for A and for B are written down, v' is eliminated by addition.

Focal length formula. The focal length f of a lens is given by

$$\frac{1}{f} = (n-1)\left(\frac{1}{r_1}+\frac{1}{r_2}\right)$$

with the "real is positive" convention, where n is the refractive index of the glass and r_1, r_2 are the radii of curvature of the respective lens surfaces.

[On the *New Cartesian* convention,

$$\left.\frac{1}{f} = (n-1)\left(\frac{1}{r_1}-\frac{1}{r_2}\right)\right].$$

Note. (1) In this formula, n is the *relative refractive index* of the medium of the lens and the medium outside the lens. Thus if a glass lens, $n = 1{\cdot}5$, is submerged in water, $n = 1{\cdot}33$, then f is calculated by using $n = 1{\cdot}5/1{\cdot}33$.

(2) The radius r_1 or r_2 is considered *positive* in sign if the surface is *convex* to the less dense medium, i.e. we observe the surface from the less dense medium. Thus with a "converging meniscus" type lens, r_1 may be $+8$ cm, $r_2 = -12$ cm, and $n = 1{\cdot}5$. Then,

$$\frac{1}{f} = (1{\cdot}5-1)\left[\frac{1}{(+8)}+\frac{1}{(-12)}\right] = +\frac{1}{48}, \quad \text{or } f = +48 \text{ cm.}$$

(2) *The power of a lens* $= 1/f$, where f is the focal length. The power of a 1 metre or 100 cm focal length lens is called *one dioptre*

(D). For any lens, with f in cm,

$$\text{number of dioptres} = \frac{100}{f}.$$

Thus +4D is a converging lens of 25 cm focal length; −2D is a diverging lens of 50 cm focal length. The greater the number of dioptres, the "stronger" is the lens.

Proof of focal length formula by small angle prism method. For small angle of incidence on a small angle prism, $d = (n-1)A$ (p. 26).

A ray parallel to the principal axis and incident at a height h above the axis on the lens is refracted towards the principal focus after passing both sides of the lens. If the two sides form part of a small angle "prism", then

$$d = h/f = (n-1)A.$$

$$\therefore \frac{1}{f} = (n-1)\frac{A}{h} \quad . \quad . \quad . \quad . \quad \text{(i)}$$

By drawing *normals* to the surfaces where the ray intersects the "prism", we find by geometry that $A = h/r_1 + h/r_2$.

$$\therefore \frac{A}{h} = \frac{1}{r_1} + \frac{1}{r_2} \quad . \quad . \quad . \quad . \quad \text{(ii)}$$

From (i) and (ii),

$$\frac{1}{f} = (n-1)\frac{1}{r_1} + \frac{1}{r_2}.$$

METHODS FOR FOCAL LENGTHS OF LENSES

Converging lens. (1) *Plane mirror method.* Coincidence of object and image gives f (see Fig. 3.3*a*,*b*).

(2) *Measurement of u and v*; then plot $1/v$ against $1/u$. The intercept on either axis gives $1/f$, from which f can be found (see p. 5) (Fig. 3.4*a*).

(3) *Magnification method.* Obtain values of m for various values of the image distance, v. Plot m against v. Since $m = v/f - 1$, the intercept on the axis of v is equal to f (Fig. 3.4*b*).

(4) *Measurement of u and v*; then plot $(u+v)$ against v. The minimum value of $(u+v)$, obtained from the graph, is equal to $4f$ (Fig. 3.4*c*).

(5) *Displacement method.* This is specially useful for a lens with inaccessible surfaces, e.g. a lens in a tube. $f = (l^2 - d^2)/4l$ (see p. 18) (Fig. 3.2).

(6) *Long focal length lens.* Combine it with a lens of moderate focal length f_1, and find the combined focal length F. Then use $1/F = 1/f_1 + 1/f_2$.

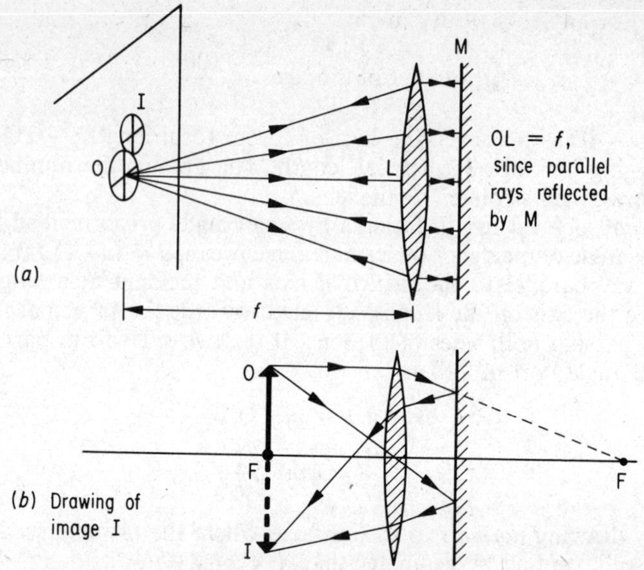

FIG. 3.3. Focal length by plane mirror.

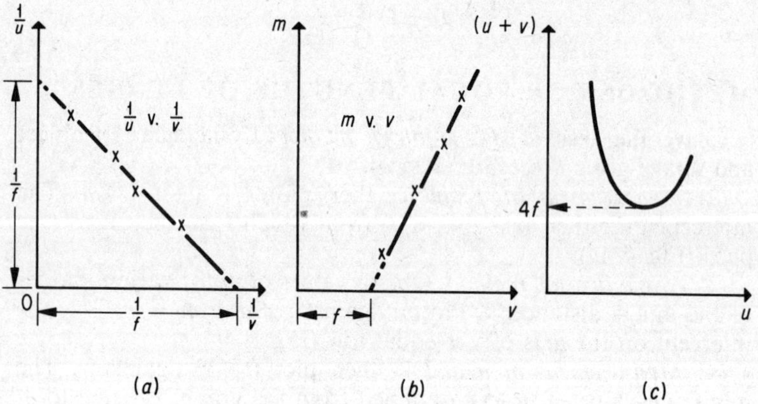

(a) Intercept on either axis = $1/f$

(b) Intercept on v–axis = f

(c) Minimum value of $(u+v) = 4f$

FIG. 3.4. Focal length by (a) $1/v$ against $1/u$ (b) magnification (c) minimum $(u+v)$.

Diverging lens. (1) *Converging lens method.* This lens M provides a *convergent* incident beam on L, or virtual object O. A real image I may then be obtained (Fig. 3.5). f can be calculated by measuring u and v and using the lens equation.

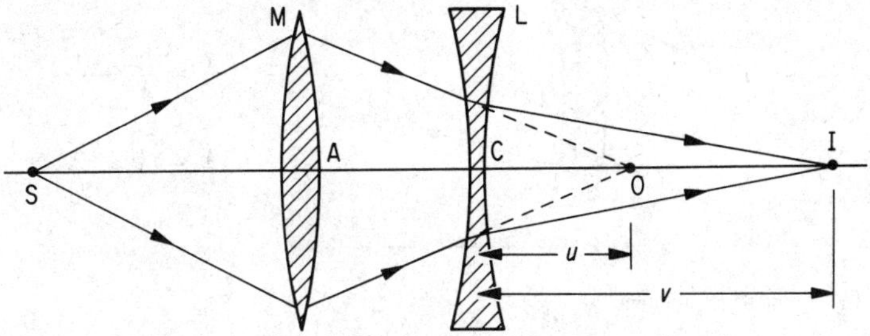

For diverging lens L, u = CO (−ve) and v = CI (+ve). Substitute in $1/v + 1/u = 1/f$

FIG. 3.5. Focal length of diverging lens.

(2) *Concave mirror method.* A concave mirror M is placed behind the diverging lens to reflect the light back (Fig. 3.6). In one position of M a clear image is obtained beside the object O. For the

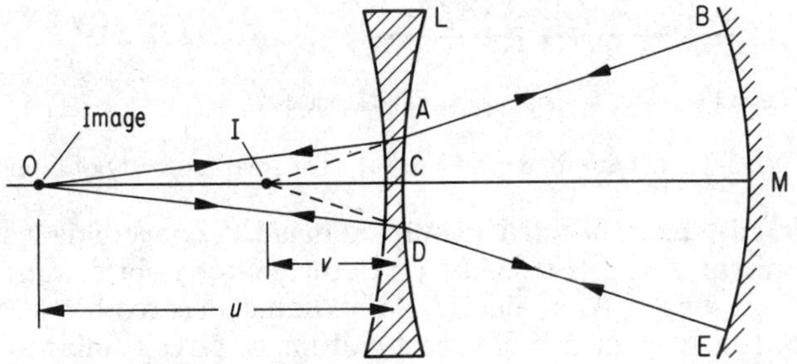

O and image coincide when M moved
Then u = OC, v = IC (−ve). Use

$$\frac{1}{v} + \frac{1}{u} = \frac{1}{f}$$

FIG. 3.6. Focal length of diverging lens.

lens, the image distance v = IC = $(r-d)$, where r is the radius of the mirror and d the distance from mirror to lens.

(3) *Converging lens combination.* With a converging lens of

greater power (shorter focal length f_1), the diverging lens may be placed in contact and the combined focal length F obtained. Then, from $1/F = 1/f_1 + 1/f_2$, the diverging lens focal length f_2 can be calculated (Fig. 3.7*a*).

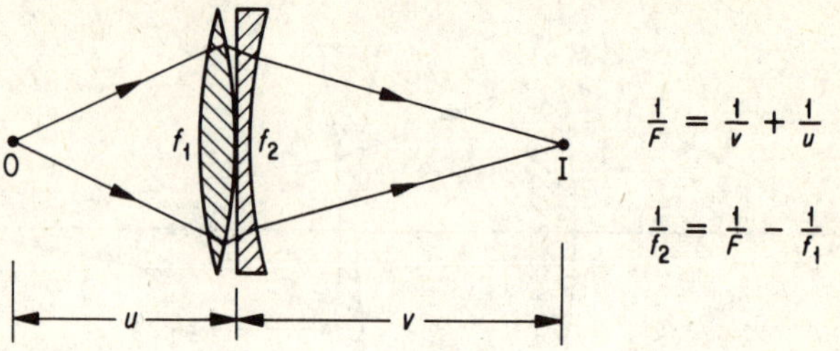

(*a*) *f* of diverging lens

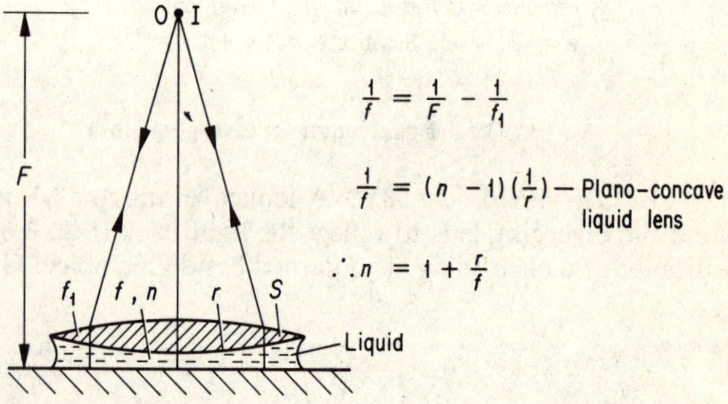

(*b*) Refractive index of small quantities of liquid

FIG. 3.7. (*a*) *f* of diverging lens (*b*) *n* for small quantity of liquid.

Refractive index of small quantity of liquid. A converging lens and plane mirror is used (Fig. 3.7*b*). (*i*) Coincidence of object and image gives focal length, f_1, of lens. (*ii*) With liquid smeared between lens and mirror, coincidence of object and image gives combined focal length, F. The liquid lens, plano-concave, has thus a focal length f given by

$$\frac{1}{F} = \frac{1}{f} + \frac{1}{f_1}.$$

The radius of curvature, r, of the underside of the lens is now found by Boys' method, and the refractive index of the liquid is calculated from the relation

$$\frac{1}{f} = (n-1)\frac{1}{r},$$

or $$n = 1+\frac{r}{f}.$$

RADII OF CURVATURE OF LENS

Converging lens. *Boys' method.* Coincidence of image and object by reflection at the back surface, of radius r. (Lens may be floated in mercury.) Then $1/r+1/u = 1/f$, from which r may be found (Fig. 3.8*a*).

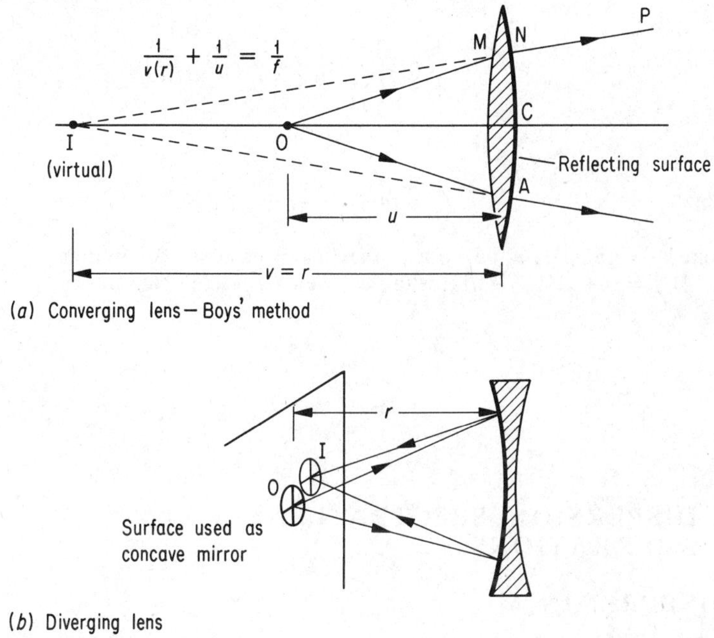

FIG. 3.8. Radius of curvature of lens.

Diverging lens. *Reflection from front surface,* which acts like a concave mirror. When object and image coincide, the distance from object to lens is r (Fig. 3.8*b*).

EXAMPLE

Explain the terms *principal foci, conjugate foci, optical centre* of a lens. Show how the radius of curvature of one face of a double convex lens may be determined by floating it on mercury.

When a thin equiconvex lens rests on a horizontal plane mirror, a pin and its image coincide in position at a point 20 cm from the lens on its principal axis. After the space between lens and mirror has been filled with a liquid the point of coincidence is found to be 7·5 cm further from the lens. If the radii of curvature of the lens surface are each 25 cm, calculate the refractive index of the liquid. (*L.*)

The focal length of the lens, f_1, = 20 cm.
The combined focal length of liquid and glass lenses, F, = 27·5 cm.
The focal length, f, of liquid lens is given by

$$\frac{1}{F} = \frac{1}{f} + \frac{1}{f_1}$$

$$\therefore \frac{1}{+27{\cdot}5} = \frac{1}{f} + \frac{1}{+20}$$

$$\therefore \frac{1}{f} = \frac{1}{27{\cdot}5} - \frac{1}{20}$$

$$\therefore f = -\frac{220}{3} \text{ cm.}$$

Now
$$\frac{1}{f} = (n-1)\frac{1}{-25} + \frac{1}{\infty},$$

from $1/f = (n-1)(1/r_2 + 1/r_2)$ with the "real is positive" convention
[$1/f = (n-1)(1/r_2 - 1/r_2)$ with the "New Cartesian" convention.]

$$\therefore \frac{-3}{220} = (n-1)\,\frac{1}{-25}$$

$$\therefore n = 1 + \frac{75}{220} = 1{\cdot}3.$$

4. DISPERSION. SPECTRA. LENS ABERRATIONS

DISPERSION

The separation of colours by a prism is called "dispersion". Small-angle prisms are studied particularly, as lenses may be considered to be built up from numerous small-angle prisms.

Deviation by small-angle prism. For a small angle of incidence, the deviation produced by a small-angle prism is given by

$$d = (n-1)A \text{ (Fig. 4.1}a\text{).}$$

Dispersion by small-angle prism. The *dispersion* between two colours, red and blue, is the *angle* between the two colours after refraction through the prism.

$$\text{Dispersion} = d_b - d_r = (n_b - n_r)A \text{ (Fig. 4.1}b\text{).}$$

Mean deviation. This is the deviation of yellow light, that of sodium light. Thus omitting the suffixes for yellow light,

$$\text{mean deviation, } d, = (n-1)A \text{ (Fig. 4.1}b\text{).}$$

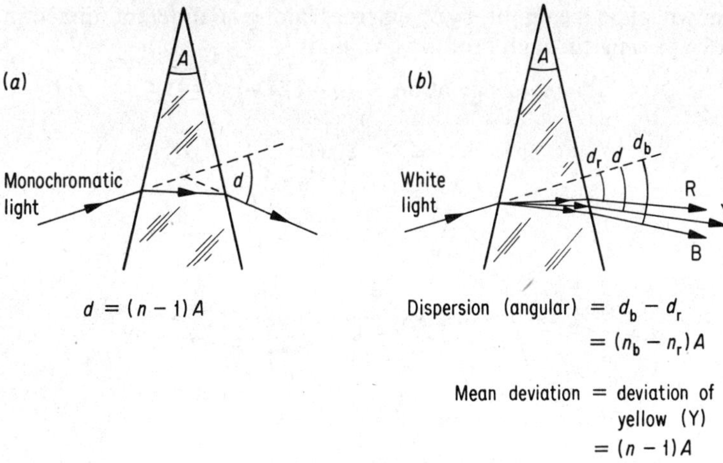

FIG. 4.1. Deviation and dispersion by small angle prism.

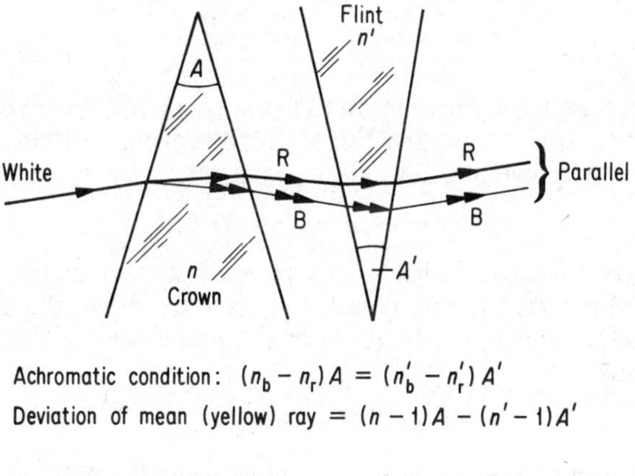

FIG. 4.2. Achromatic prisms.

Achromatic prisms. These consist of a crown glass prism and a flint glass prism, with one of the prisms inverted (Fig. 4.2). To eliminate the dispersion between the blue and red colours in white light,

$$(n_b - n_r)A = (n'_b - n'_r)A';$$

the two sets of symbols refer to crown glass and flint glass respectively. This formula enables the angle of one prism to be found when the angle of the second prism, and the refractive indices, are known. The beam of light emerges along a different direction after passing through both prisms, and

$$\text{the mean deviation} = (n-1)A-(n'-1)A'.$$

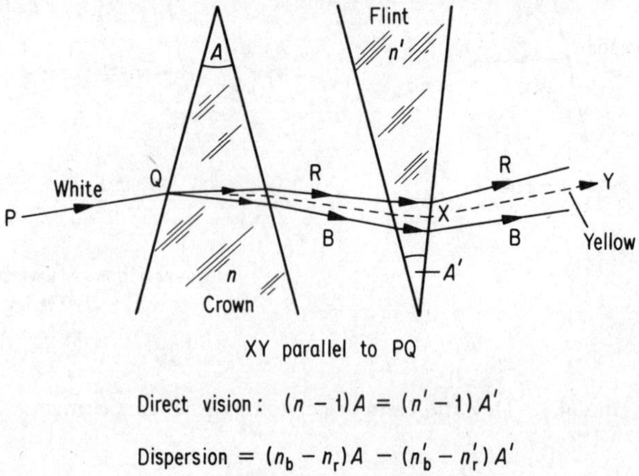

FIG. 4.3. Direct vision spectroscope.

Direct-vision spectroscope. In this case there is no mean deviation, and dispersion occurs. See Fig. 4.3. The former condition means that, with a crown and flint glass prism,

$$(n-1)A = (n'-1)A',$$

where the refractive index is that for yellow (sodium) light. Since one of the prisms is inverted with respect to the other, the dispersion between the blue and red light after refraction through both prisms is given by

$$d = (n_b-n_r)A-(n'_b-n'_r)A'.$$

Dispersive power. The *dispersive power,* ω, of a material for blue and red light is defined by the relation

$$\omega = \frac{n_b-n_r}{n-1}.$$

For a small-angle prism, ω is defined by

$$\omega = \frac{\text{angular dispersion between blue and red}}{\text{mean deviation}}$$

$$= \frac{d_b-d_r}{d} = \frac{(n_b-n_r)A}{(n-1)A} = \frac{n_b-n_r}{n-1}.$$

Measurement of dispersive power. A sodium flame is placed in front of the collimator slit of an adjusted spectrometer, with a prism of the material on the table, and the angle of minimum deviation determined (Fig. 4.4). The flame is now replaced by a

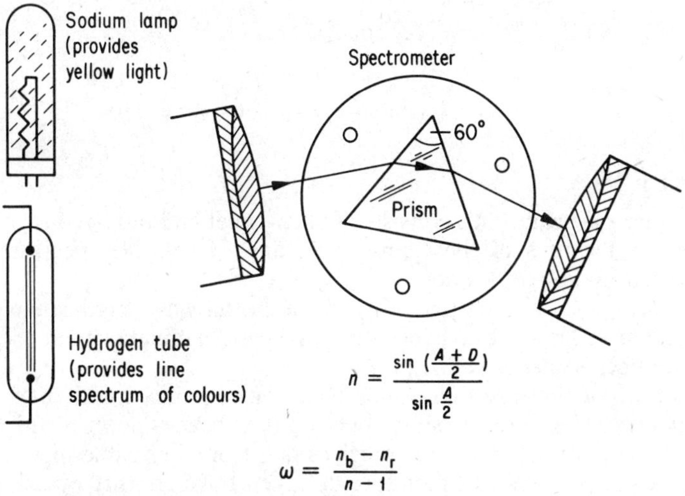

FIG. 4.4. Measurement of dispersive power.

glowing hydrogen tube, and minimum deviation for the red and blue lines in the spectrum are now measured. n, n_b, n_r are each found from $n = \sin(A+D/2) \div \sin(A/2)$, and ω then calculated from $\omega = (n_b - n_r)/(n-1)$.

EXAMPLE

For a particular flint glass, red, blue and yellow light have respective refractive indices of 1·644, 1·650, 1·664; for a particular crown glass, the same colours have respective refractive indices of 1·514, 1·517, 1·523. Calculate (*i*) the dispersive power of flint and crown glass for the red and blue light, (*ii*) the refracting angle of a crown glass prism which produces an achromatic combination with a 4° flint glass prism and the mean deviation of light in this case.

(*i*) $$\omega = (n_b - n_r)/(n-1)$$

Flint glass: $$\omega = \frac{1{\cdot}664 - 1{\cdot}644}{1{\cdot}650 - 1} = 0{\cdot}0308$$

Crown glass: $$\omega = \frac{1{\cdot}523 - 1{\cdot}514}{1{\cdot}517 - 1} = 0{\cdot}174$$

(*ii*) For achromatic combination,

$$(n_b - n_r)A = (n'_b - n'_r)A'$$

$$\therefore (1{\cdot}523 - 1{\cdot}514)A = (1{\cdot}664 - 1{\cdot}644)4^\circ$$

$$\therefore A = \frac{0{\cdot}02 \times 4}{0{\cdot}009} = 9^\circ$$

$$\text{Mean deviation} = \text{deviation of yellow light}$$

$$= (n-1)A - (n'-1)A'$$

$$= (1{\cdot}517 - 1)9 - (1{\cdot}650 - 1)4 = 2{\cdot}05^\circ.$$

SPECTRA

Solar spectrum. This consists of ultra-violet and infra-red rays, as well as the rays of the visible spectrum. The visible spectrum is crossed by fine dark lines.

Ultra-violet rays are rays which have shorter wavelengths than the violet rays. They are invisible, but cause certain minerals to fluoresce and affect a photographic plate.

Infra-red rays are rays which have longer wavelengths than the red rays. They are invisible, but produce heat. The rays are not dispersed by mist and fog as much as rays from the visible spectrum, and clear pictures can therefore be taken through mist by using a camera with an infra-red filter and an infra-red sensitive photographic plate.

Visible spectrum. These rays cause the sensation of vision. The colours range through red, orange, yellow, green, blue, indigo, to violet.

Fine dark lines crossing solar spectrum. These are called Fraunhofer lines, after the discoverer, and indicate the presence of certain elements such as hydrogen and sodium in the sun's atmosphere. These elements scatter the light from the sun which the elements would themselves emit when hot. The light passing through the elements is therefore deficient in these wavelengths, and there is a shadow on the corresponding part of the solar spectrum. See Fig. 4.6*b*.

Production of pure spectrum. A source of white light such as an arc light is placed in front of the narrow slit of a collimator of an adjusted spectrometer (p. 15), and a glass prism is placed on the table. On looking through the telescope a pure spectrum is received.

With only one lens, the narrow slit is placed so that a real image I of it is obtained, and the prism is placed beyond the lens in the minimum deviation position. A fairly pure spectrum is then obtained on a screen beyond the prism near I.

Detection of ultra-violet and infra-red light. Ordinary glass absorbs most of the ultra-violet and infra-red rays. A narrow slit, a quartz lens, a quartz prism and a screen in a dark room produce ultra-violet rays. Fluorite or rocksalt for infra-red rays. To detect the latter a thermopile or phototransistor connected to a sensitive galvanometer is moved into the dark portion beyond the red, when a deflection is obtained. The ultra-violet rays are detected by moving a tube of fluorescein or a photocell sensitive to ultra-violet into the dark portion beyond the violet end. The figures on a luminous watch-dial also glow in this part of the spectrum.

Emission spectra. These divide into three main classes:

(*i*) *Line spectra.* Using a prism and spectrometer, a series of lines of different wavelength are obtained when the atoms of a hot glowing substance are observed (Fig. 4.5*a*). Vaporized sodium gives two prominent yellow lines close to each other; glowing hydrogen or helium gas give lines of different colours.

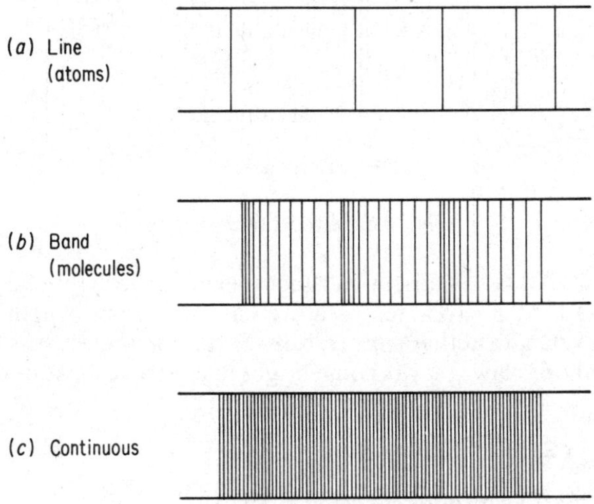

FIG. 4.5. Emission spectra.

(*ii*) *Band spectra.* These consist of a series of fluted bands, each sharp at one end but fading at the other. The bands are actually a series of numerous fine lines close to each other (Fig. 4.5*b*). Oxygen and nitrogen give band spectra, which are usually emitted by molecules.

(*iii*) *Continuous spectra.* These are obtained from solids and liquids. The sun's spectrum is continuous (Fig. 4.5*c*).

Absorption spectra. When a sodium flame is placed in front of the continuous spectrum emitted from an arc for example, the spectrum obtained is crossed by a dark line. This is an example of an absorption spectrum. The sodium absorbs yellow light from the white light and re-radiates it in all directions, so that the yellow light beyond the flame is less intense (Fig. 4.6*a*).

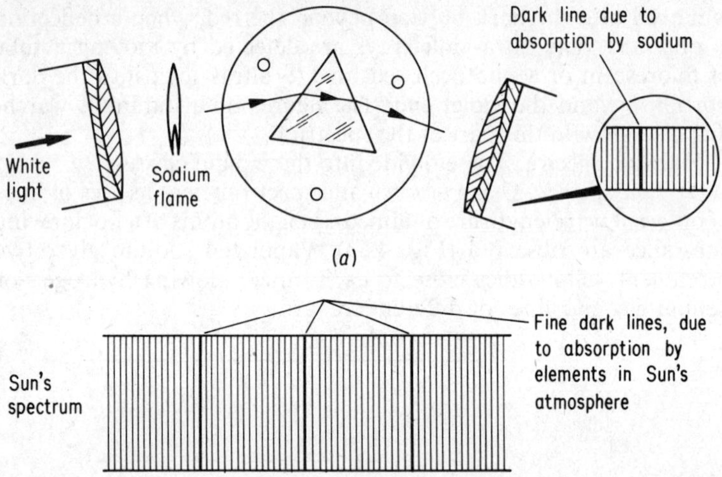

(*b*) Fraunhofer lines

FIG. 4.6. Absorption spectra.

Kirchhoff's law states: A substance which emits light of a certain wavelength at a given temperature can also absorb light of the same wavelength at that temperature. Fraunhofer lines are examples of Kirchhoff's law; they become bright lines at an eclipse of the sun.

LENS ABERRATIONS

Focal length with different colours. Since $1/f = (n-1)(1/r_1 + 1/r_2)$, and the refractive index of a material varies with colour, f varies with different colours. Now the refractive index of glass for blue light, n_b, is greater than that for red light, n_r. Hence from the formula, the focal length for blue light, f_b, is less than that for red light, f_r (Fig. 4.7*a*).

If one focal length is known, the other can be found by eliminating the radii of curvature from the two formula. Then

$$\frac{f_r}{f_b} = \frac{n_b - 1}{n_r - 1}.$$

Chromatic aberration. Achromatic lenses. Since the focal length of a lens varies with the colour of the light, the image of an object in a lens is coloured. This defect of a lens is known as *chromatic aberration.*

Two lenses in contact which bring the blue and red rays to the same focus are called an *achromatic doublet* for these colours. They obey the relation

$$\frac{f_1}{f_2} = -\frac{\omega_1}{\omega_2}, \qquad \text{(i)}$$

where f_1, f_2 are the focal lengths and ω_1, ω_2 are the respective dispersive powers $[(n_b - n_r)/(n-1)]$ of the glasses (see proof later). A telescope objective is usually made of a biconvex crown glass lens and a plano-concave flint glass lens, cemented together; the incident light falls on the converging or convex lens (Fig. 4.7*b*).

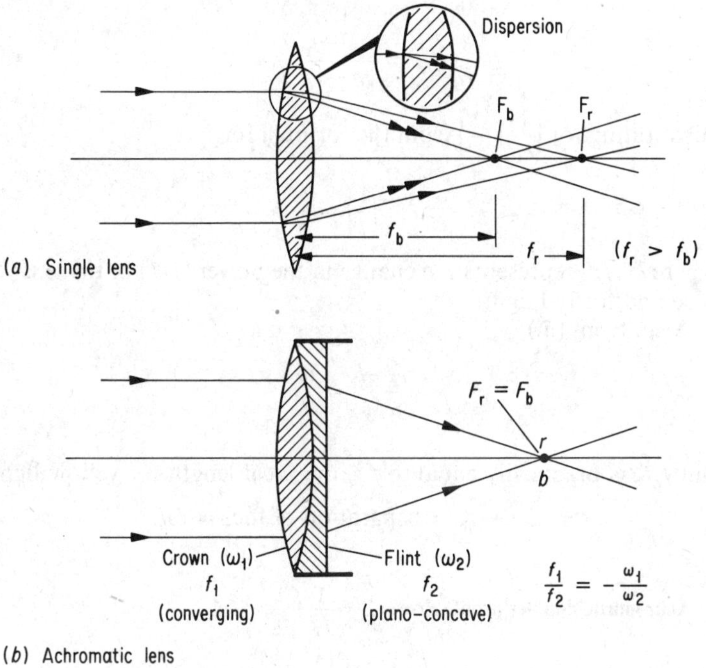

FIG. 4.7. (*a*) Chromatic aberration (*b*) Achromatic doublet.

If the objective is required to have a focal length F of 100 cm, then

$$\frac{1}{f_1} + \frac{1}{f_2} = \frac{1}{F} = \frac{1}{100} \qquad \text{(ii)}$$

Knowing ω_1, ω_2, F, the focal lengths f_1, f_2 can each be calculated from (i) and (ii). Then, since $1/f_2 = (n_2 - 1)(1/r)$ for the plano-concave lens, the radius of curvature r can be found. The radius of curvature of the second face of the convex lens can now be found from the relation

$$\frac{1}{f_1} = (n_1-1)\left(\frac{1}{r_1}+\frac{1}{r_2}\right).$$

Separation of foci. Since

$$\frac{1}{f} = (n-1)\left(\frac{1}{r_1}+\frac{1}{r_2}\right),$$

then

$$\frac{1}{f_b}-\frac{1}{f_r} = (n_b-n_r)\left(\frac{1}{r_1}+\frac{1}{r_2}\right)$$

$$\therefore \frac{1}{f_b}-\frac{1}{f_r} = \frac{n_b-n_r}{n-1}\cdot\frac{1}{f}, \quad . \quad . \quad . \quad . \quad \text{(iii)}$$

substituting for $\left(\frac{1}{r_1}+\frac{1}{r_2}\right)$ from the formula for f.

$$\therefore \delta\left(\frac{1}{f}\right) = \frac{\omega}{f},$$

where $\delta(1/f)$ represents the change in the power $(1/f)$ of the lens for blue and for red light.

Also, from (iii),

$$\frac{f_r-f_b}{f_r f_b} = \frac{\omega}{f}.$$

But $f_r f_b$ is practically equal to f^2, the focal length for yellow light.

$$\therefore f_r-f_b = \text{separation of foci} = \omega f.$$

Achromatic doublet proof. From $\frac{1}{F} = \frac{1}{f_1}+\frac{1}{f_2}$,

we have

$$\delta\left(\frac{1}{F}\right) = \delta\left(\frac{1}{f_1}\right)+\delta\left(\frac{1}{f_2}\right)$$

$$\therefore \delta\left(\frac{1}{F}\right) = \frac{\omega_1}{f_1}+\frac{\omega_2}{f_2}.$$

For achromatic doublet, $\delta\left(\frac{1}{F}\right) = 0.$

$$\therefore \frac{\omega_1}{f_1}+\frac{\omega_2}{f_2} = 0,$$

or

$$\frac{f_1}{f_2} = -\frac{\omega_1}{\omega_2}.$$

Spherical aberration. This occurs when a wide beam of light falls on a lens; rays at O further from the axis are brought to a nearer focus F_0 than rays close to the principal axis (Fig. 4.8*a*). The aberration can be reduced by having a diaphragm which allows central rays to be incident on the lens; this reduces the brightness of the image since some light is lost (Fig. 4.8*b*). A better method is to reduce the angle of incidence on a particular lens surface, by sharing equally the total deviation of the light between the lens surfaces (Fig. 4.8*b*).

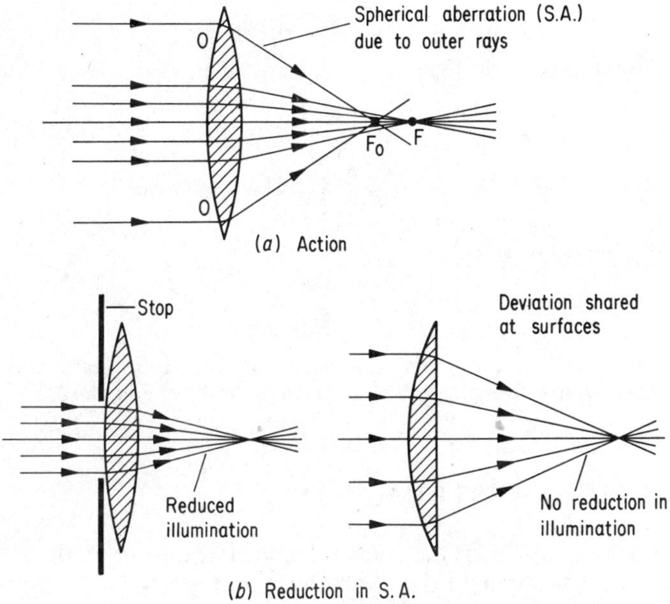

FIG. 4.8. Spherical aberration and reduction.

For a single lens, spherical aberration is reduced when parallel light is incident on the curved surface of a *plano-convex* lens. Two lens reduce spherical aberration considerably when the distance between them $= f_1 - f_2$, where f_1, f_2 are their respective focal lengths.

EXAMPLE

It is required to construct a telescope object glass, of focal length 100 cm, which is to be achromatic for the C and F lines of the spectrum. Using the data below, find the nature and focal lengths of the two thin lenses of crown and flint glass respectively which, placed in contact, will give the desired result. Derive the necessary formula. (Crown glass, $n_C = 1{\cdot}513$, $n_F = 1{\cdot}521$; flint glass, $n_C = 1{\cdot}604$, $n_F = 1{\cdot}620$.)

Dispersive power of crown glass, ω_1, $= \frac{1{\cdot}521-1{\cdot}513}{1{\cdot}517-1} = \frac{8}{517}$ assuming n_D (yellow) is the average of n_C and n_F.

Dispersive power of flint glass, ω_2, $= \frac{1{\cdot}620-1{\cdot}604}{1{\cdot}612-1} = \frac{16}{612} = \frac{4}{153}$.

If f_1, f_2 are the corresponding focal lengths, then

$$\frac{f_1}{f_2} = -\frac{\omega_1}{\omega_2} = -\frac{8}{517}\times\frac{153}{4} = -\frac{306}{517} \quad . \quad . \quad . \quad . \quad . \quad \text{(i)}$$

Also,
$$\frac{1}{f_1}+\frac{1}{f_2} = \frac{1}{F} = \frac{1}{+100} \quad . \quad . \quad . \quad . \quad . \quad . \quad . \quad . \quad \text{(ii)}$$

Substituting for f_2 for (i) in (ii),

$$\therefore\ -\frac{517}{306}.\frac{1}{f_2}+\frac{1}{f_2} = \frac{1}{100}$$

$$\therefore f_2 = \frac{211\times 100}{306} = -69{\cdot}0 \text{ cm.}$$

Substituting for f_2 in (i),

$$f_1 = +\frac{306}{517}\times 69 = +40{\cdot}8 \text{ cm.}$$

Thus a converging lens of focal length 40·8 cm, crown glass, must be combined with a diverging lens of focal length 69·0 cm, flint glass.

5. OPTICAL INSTRUMENTS

Visual angle. This is the angle subtended at the eye by the object or image. The apparent size of an object is proportional to its visual angle, and in the theory of optical instruments we are therefore primarily concerned with visual angles and not with the actual lengths of the objects or images.

Microscope. The *angular magnification* or *magnifying power, M,* is defined by

$$M = \frac{\alpha'}{\alpha},$$

where α' is the visual angle of the final image with the microscope, usually at the least distance of distinct vision, and α is the visual angle of the object *placed at the least distance of distinct vision* and viewed by the unaided eye (Fig. 5.1*a*, *b*).

Simple microscope, or magnifying glass. Since $M = \alpha'/\alpha$, and $\alpha' = h'/D$, $\alpha = h/D$, then $M = h'/h$, where h' is the length of the image and h is the length of the object. Thus M is also the linear magnification produced by the lens. Generally, from p. 18,

$$\text{linear magnification} = \frac{v}{f} - 1$$

$$\therefore M = -\frac{D}{f} - 1,$$

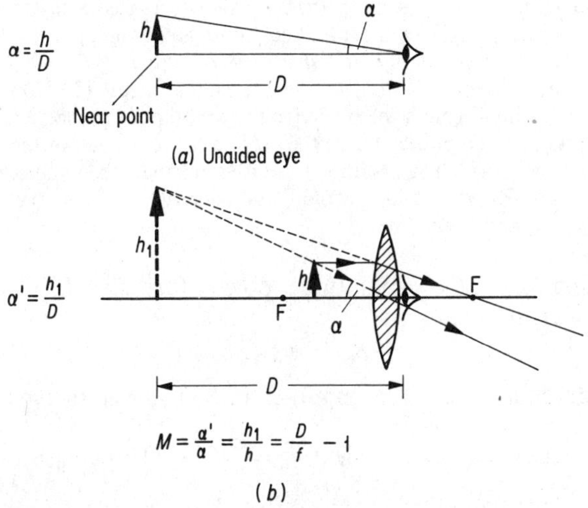

FIG. 5.1. Angular magnification of microscope.

where D is *numerically* the least distance of distinct vision (Fig. 5.1*b*).

[*New Cartesian convention.* Since linear magnification $= 1 - v/f$, $\therefore M = 1 + \dfrac{D}{f}$.]

Compound microscope. This has two converging lenses, both of short focal length (Fig. 5.2). In using the instrument,

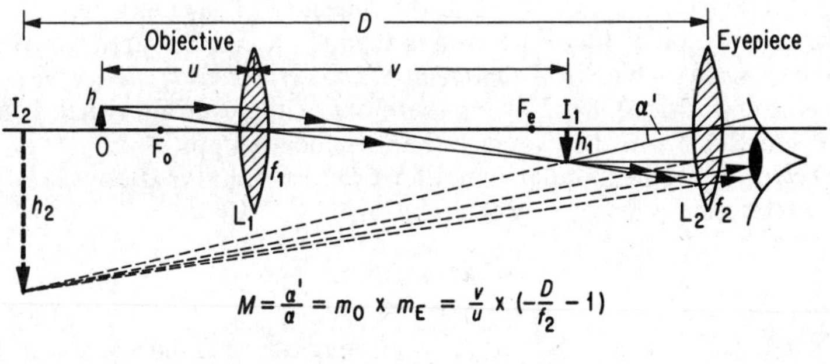

FIG. 5.2. Compound microscope.

(*i*) the object is placed slightly *further* from the objective than its focal length,

(*ii*) the real image formed is viewed through the eyepiece, acting as a magnifying glass.

To draw the rays which enter the eye: (1) Use I_1 as an object for the eyepiece. (2) From the top of I_1 draw (*i*) a line through the lens L_2 centre and (*ii*) another line parallel to the principal axis which passes to the principal focus (not shown). Produce the lines back to form the top point of the virtual image I_2, and note carefully that *these are construction lines—not rays of light*—so that "arrows" must not be drawn on them. (3) Now produce those rays passing through the objective to the top of I_1 to meet the eyepiece lens L_2, and join the points where they strike the lens back to the top of I_2 by dotted lines. (4) Then produce the latter through the eyepiece, and as these are actual rays passing through the eyepiece, draw in the eye-pupil so that the rays enter the eye.

Starting from the magnifying power formula, $M = \alpha'/\alpha$, we arrive at the result

$$M = m_o \times m_E,$$

where m_o, m_E are the *linear* magnifications produced by the objective and eyepiece respectively. Since $m_o = (v/f_1 - 1)$ and $m_E = \left(-\frac{D}{f_2} - 1\right)$, where v is the real image distance from the objective, f is the objective focal length, and f_2 is the eyepiece focal length,

$$M = \left(\frac{v}{f_1} - 1\right)\left(-\frac{D}{f_2} - 1\right).$$

[*New Cartesian convention.* Since $m_o = 1 - v/f_1$ and $m_E = 1 + D/f_2$,

$$\therefore M = \left(1 - \frac{v}{f_1}\right)\left(1 + \frac{D}{f_2}\right)\Big].$$

Best position of eye. The eyering. All the rays which enter the optical instrument pass through the objective. The emergent beam from the eyepiece therefore passes through a circular area known as the *eyering*, which is *the image of the objective lens in the eyepiece*. The best position of the eye is therefore at the eyering, which is a little further from the eyepiece than its focal length.

Telescopes. The angular magnification or magnifying power, M, of a telescope is defined by the relation

$$M = \frac{\alpha'}{\alpha},$$

where α' is the visual angle of the final image and α is the visual angle of the distant object viewed with the unaided eye. See Fig. 5.3*a*.

Note that an objective of large diameter (*i*) collects more light from the distant object so the image is brighter, (*ii*) increases the resolving power (p. 73).

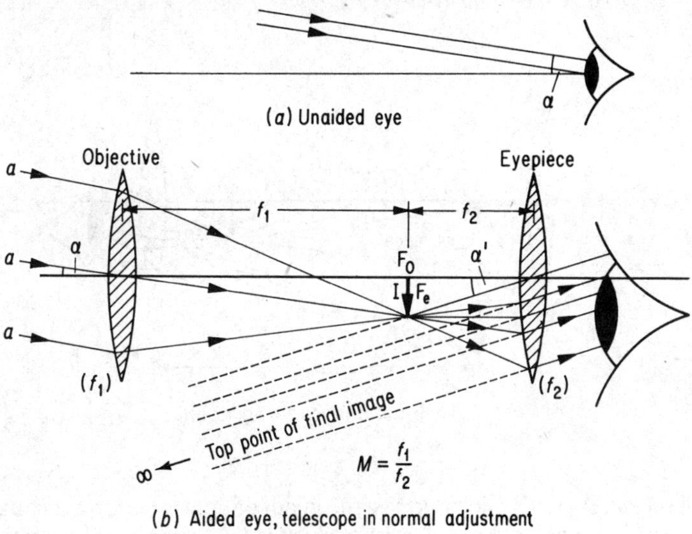

FIG. 5.3. Angular magnification of telescope.

Astronomical Telescope. Final image at infinity. In this case the telescope is said technically to be in *normal adjustment* (Fig. 5.3*b*). The following points may be noted:

(*i*) In drawing, take three parallel rays incident on the objective from the distant *top a* of the object. Further, these rays make a small angle α with the principal axis.

(*ii*) The rays emerge parallel through the eyepiece.

(*iii*) In the drawing, note the construction lines (without arrows) to the eyepiece from the top point of I, and the *actual rays* from the latter to the eyepiece, which are continuous from the objective and which emerge from the eyepiece and enter the eye.

(*iv*) The angle α is practically equal to the visual angle of the distant object, since the length of the telescope is negligible compared with the distance of the object from the eye (Fig. 5.3*b*).

(*v*) The magnifying power, M, $= \dfrac{f_1}{f_2}$,

where f_1, f_2 are the focal lengths of objective and eyepiece respectively.

(*vi*) The distance between the lenses $= f_1 + f_2$.

Astronomical telescope. Final image at near point. The final image is now virtual and distant D from the eyepiece; the image produced by the objective is thus nearer to the eyepiece than its focal length (Fig. 5.4).

$$\text{Magnifying power, } M, = \frac{f_1}{f_2}\left(1+\frac{f_2}{D}\right).$$

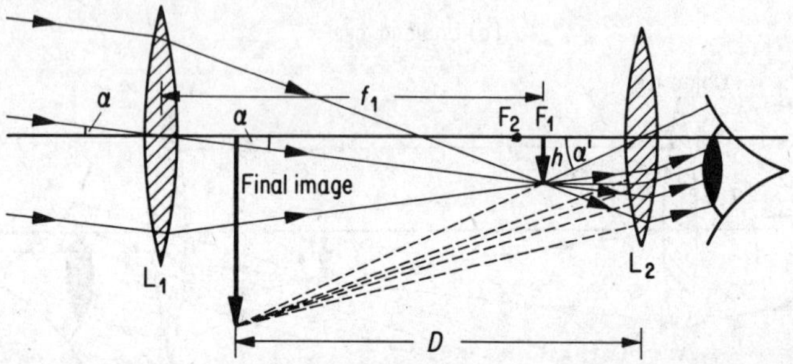

FIG. 5.4. Final image at near point.

Eyering. As with microscopes, the beam of light emerging through the eyepiece passes through a ring *which is the image of the objective in the eyepiece*. This ring is called the eyering, and is the best position for the eye. Since the eyering is real, the astronomical telescope has a wide field of view. See Fig. 5.7*a*, p. 43.

With the telescope in normal adjustment,

$$\text{magnifying power, } M, = \frac{\text{diameter of objective}}{\text{diameter of eyering}}.$$

Proof: Diameter of objective:diameter of eyering $= u:v = (f_1+f_2):v$, where v is the image distance when $u = f_1+f_2$ and $f = f_2$. Substitute in $1/v+1/u = 1/f$ to find v, and evaluate $(f_1+f_2):v$.

Terrestrial telescope. Astronomical telescopes give a final *inverted* image. The terrestrial telescope gives a final erect image. It has a converging objective (f_1), a converging erecting lens (f) at a distance (f_1+2f) from the objective, and a converging eyepiece (f_2) at a distance (f_2+2f) from the erecting lens when the final image is at infinity.

The total length of the telescope is (f_1+f_2+4f).

The magnifying power is f_1/f_2.

Loss of light by reflection at the lens surfaces, and its long length, are two disadvantages.

Galilean telescope. Provides an erect final image with two lenses only. The objective is a converging lens of long focal length f_1; the eyepiece is a diverging lens of short focal length f_2; with final image at infinity, the distance apart of the lenses is numerically $(f_1 - f_2)$, the difference in the focal lengths.(Note that the distance apart in the astronomical telescope in normal adjustment is the sum of the focal lengths.)

The magnifying power, M, $= f_1/f_2$. The image formed by the objective would be at a point *beyond* the eyepiece, but the latter intercepts the convergent beam and it emerges parallel.

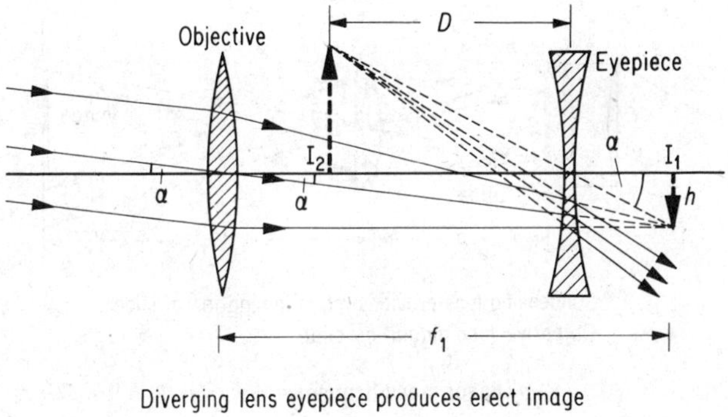

FIG. 5.5. Galilean telescope; final image at near point.

Advantage and disadvantage. The telescope provides an erect final image with two lenses only (Fig. 5.5), and it is a short telescope compared with the terrestrial one. It is therefore used in *opera glasses*. Since the image of the objective in the diverging eyepiece is virtual, the eyering is virtual. Fig. 5.7*b*, p. 43. The best position of the eye is thus as close to the eyepiece as possible, and the field of view is restricted. This is the main disadvantage of the Galilean telescope.

Prism binoculars. Has a converging objective and a converging eyepiece, with two right-angle isosceles (total reflecting) prisms between them. The optical path of the light = (objective to prism A) + (prism A to prism B) + (prism B to eyepiece) = 3 × geometrical length (approx.) between objective and eyepiece.

Prism A inverts the image laterally; prism B inverts it vertically. The final image is thus erect. The advantage of the prism binoculars is (*i*) a wide field of view, since the image of the objective in the converging eyepiece is real, (*ii*) it is a short telescope, providing the same magnification as an astronomical telescope about three times as long, (*iii*) binocular vision.

Reflector telescopes. Loss of light mainly by reflection at lens' surfaces makes lenses unsuitable for large-scale astronomy. Newton was the first person to suggest the use of a large *mirror* to collect the light from distant stars and planets. The Mount Palomar telescope has a 5 metre wide parabolic mirror, with smaller plane and convex mirrors to bring the light to a more convenient focus. The parabolic mirror is coated on the front with aluminium.

Projection lantern. This has (*i*) a powerful point source of light O, (*ii*) a condensing lens (2 plano-convex lenses) which concentrate the light towards the slide, (*iii*) an objective lens, achromatic, which is

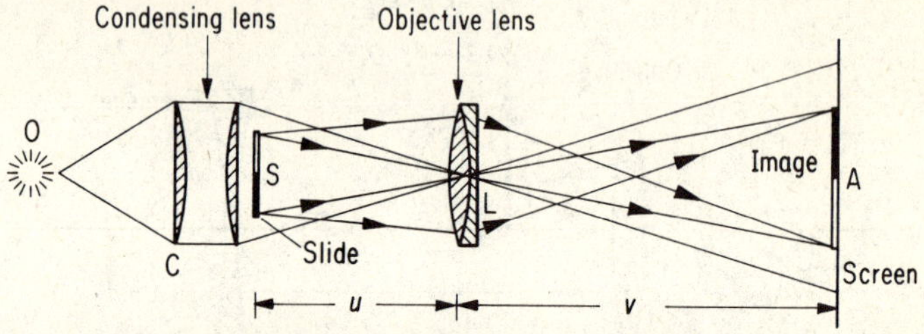

Condensing lens provides high illumination of slide
Objective lens magnifies slide

m (linear magnification) = $\frac{v}{u}$

FIG. 5.6. Projection lantern.

adjusted so that a clear image is focussed on the screen. The objective lens is at the position of the image of O in the condensing lens, so that a lamp image does not appear on the screen (Fig. 5.6).

The linear magnification of the slide, $\frac{v}{u}$, $= \frac{v}{f} - 1$.

[*New Cartesian convention.* Linear magnification of slide $= 1 - v/f$.]

Hence high magnification is obtained with an objective whose focal length is small compared with the distance of the screen from the lens. Note that we are concerned here with "linear" magnification, *not* "angular" magnification.

Lens camera. Depth of field. The lens is corrected for chromatic aberration and spherical aberration. The exposure time depends on the f-number, which affects the brightness of the image (see p. 52).

The eye can see as a reasonably sharp point a circle of diameter about 0·25 mm, the "circle of least confusion". Thus a point object O, photographed by a camera lens, will be clearly seen on developing the film, if its image forms a *circle* not greater than the circle of least confusion. The range of distance of O over which it can be clearly photographed is called the *depth of field*. Generally, the depth of field increases as the lens aperture decreases.

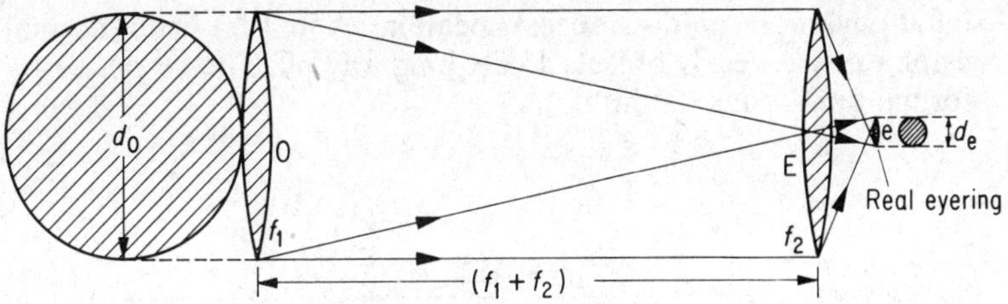

Wide field of view due to real (accessible) eyering

$$M = \frac{d_o}{d_e}$$

(*a*) Astronomical telescope, normal adjustment

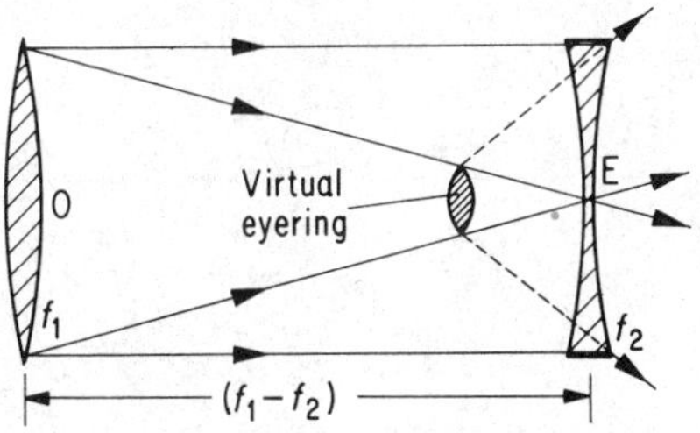

Limited field of view due to virtual eyering
Prism binoculars has wide field of view (converging lens eyepiece)

(*b*) Galilean telescope, normal adjustment

FIG. 5.7. Eyering in (*a*) astronomical, (*b*) Galilean telescope.

6. DEFECTS OF VISION. VELOCITY OF LIGHT. PHOTOMETRY

DEFECTS OF VISION

The eye. This contains (*i*) a crystalline lens, (*ii*) ciliary muscles which control the radii of the lens surfaces and hence the focal length, (*iii*) the retina, a sensitive "screen" at the back of the eyeball, covered with nerves linked to the optic nerve. The iris of the eye is a diaphragm with a circular (black) hole in the middle called the pupil; the colour of an eye is the colour of the iris. The aqueous humour is a weak salt solution in front of the lens; the vitreous humour is a gelatinous substance between the lens and retina. The cornea is the tough transparent spherical bulge in front of the lens.

The use of two eyes gives a three-dimensional view, bringing with it a sense of distance.

Far and near points. Accommodation. A person with normal vision can see clearly objects a very long way off. The *far point* of a normal eye is thus "infinity".

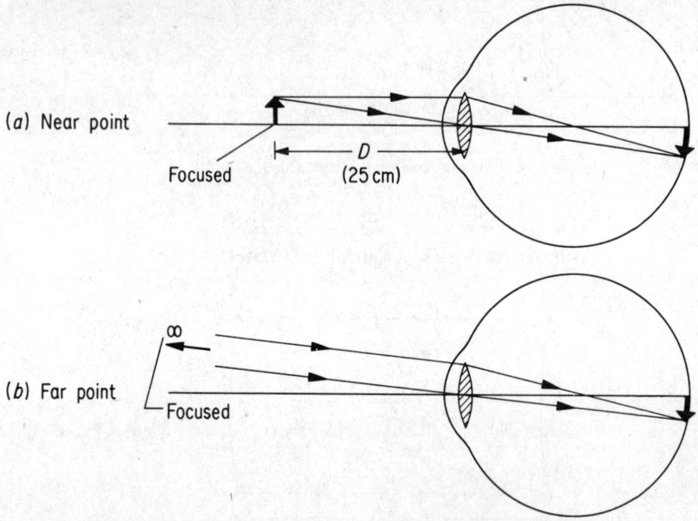

FIG. 6.1. Far and near points in normal vision.

The nearest point which the eye can see distinctly is about 25 cm for normal vision. The distance concerned, *D*, is called *the least distance of distinct vision* of the eye. See Fig. 6.1*a*,*b*.

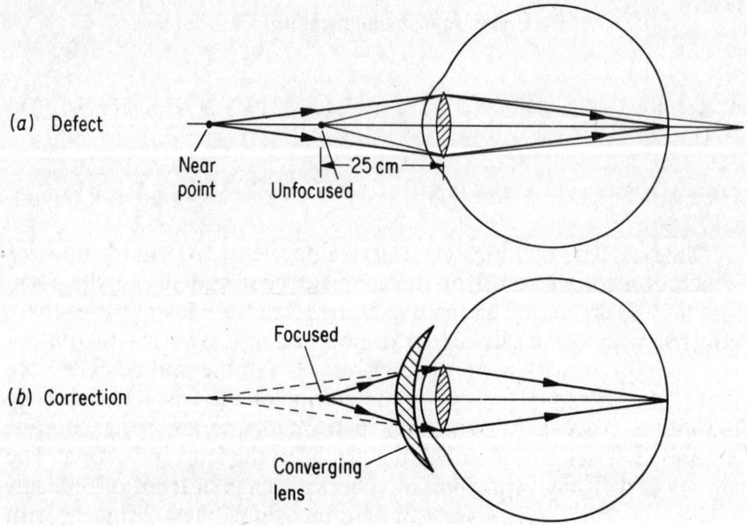

FIG. 6.2. Far sight and correction.

With normal vision, the eye is fully relaxed or "unaccommodated" when viewing objects at infinity. When the object is at the least distance of distinct vision and is focused by the eye, the ciliary muscles are fully strained and the eye is said to be "fully accommodated".

Far sight. In this defect of vision the far point of the eye is normal, but the near point is further from the eye than 25 cm (Fig. 6.2*a*). Correction is made by a suitable *converging* lens (Fig. 6.2*b*).

Short sight. In this defect the far point is near to the eye, and rays from infinity now come to a focus in front of the retina (Fig. 6.3*a*).

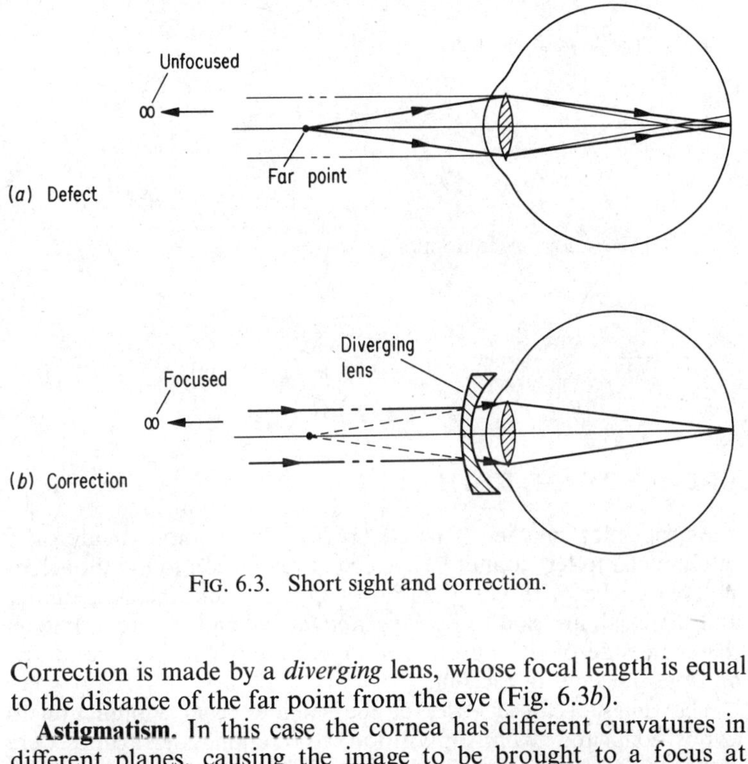

FIG. 6.3. Short sight and correction.

Correction is made by a *diverging* lens, whose focal length is equal to the distance of the far point from the eye (Fig. 6.3*b*).

Astigmatism. In this case the cornea has different curvatures in different planes, causing the image to be brought to a focus at different places. Correction is made by a suitable *cylindrical* lens.

If we wish to correct for short sight, when the far point is 300 cm away for example, the lens must refract parallel rays so that they appear *after refraction* to come from a point 300 cm away. (A suitable diverging lens is needed.) Similarly, if in far sight the near point is 50 cm away, the correction lens must refract rays coming from a point 25 cm away so that they appear *after refraction* to come from a point 50 cm away. (A suitable converging lens is needed.)

EXAMPLE

The range of distinct vision of a short-sighted person extends between the distances of 10 cm and 30 cm from the eye. Calculate the focal length of the spectacle lenses which he must wear to see distant objects at his far point. With these spectacles what would be the least distance of his distinct vision?

To see distant objects clearly, parallel rays must appear to come from a point 30 cm from the eye after passing through the lens. Thus a *concave* or *diverging* lens of focal length 30 cm must be used.

With these lenses, an object at the new least distance of distinct vision (u) must give rise to rays which appear to come from a point 10 cm away after refraction. Applying the lens equation,

with the "real is positive" convention, $\frac{1}{v}+\frac{1}{u} = \frac{1}{f}$,

$$\frac{1}{-10}+\frac{1}{u} = \frac{1}{-30}$$

$$\therefore u = 15 \text{ cm.}$$

[*New Cartesian* convention, $\frac{1}{v}-\frac{1}{u} = \frac{1}{f}$

$$\therefore \frac{1}{-10}-\frac{1}{u} = \frac{1}{-30}$$

$$\therefore u = -15 \text{ cm.}]$$

VELOCITY OF LIGHT

Astronomical method (Romer) (1676). Observations made on a satellite of Jupiter. Jupiter has period of revolution round the sun of $11\frac{3}{4}$ years; the earth has a period of 1 year round the sun. Jupiter and the earth are said to be *in conjunction* when they are in line on the same side of the sun, and *in opposition* when they are in line on opposite sides of the sun.

The times of the eclipses of the satellite from conjunction to opposition, and from opposition to conjunction again, were recorded. Although the number of eclipses was the same in each case, the last eclipse was observed to be $16\frac{1}{2}$ minutes late from conjunction to opposition, and $16\frac{1}{2}$ minutes too early from opposition to conjunction. *Römer concluded that light took* $16\frac{1}{2}$ *minutes to travel across the diameter of the earth's orbit round the sun,* about 300 million kilometres·

$$\therefore \text{velocity of light} = \frac{300 \times 10^6}{990} = 3 \times 10^5 \text{ km s}^{-1} \text{ (approx.)}$$

Terrestrial method. Rotating wheel (Fizeau) (1849). Apparatus used was:

(*i*) Rotating toothed wheel W (720 teeth),

(*ii*) a plane glass F which reflected light from a powerful source of light to the rim of the wheel,

(*iii*) a converging lens B placed at a distance equal to its focal length from the wheel, so that parallel light travelled through the lens,

(*iv*) a second converging lens C a long way from the wheel which brought the light to a focus on a plane mirror beyond it,

(*v*) a telescope sighting through the plane glass, and adjusted for parallel light (see Fig. 6.4).

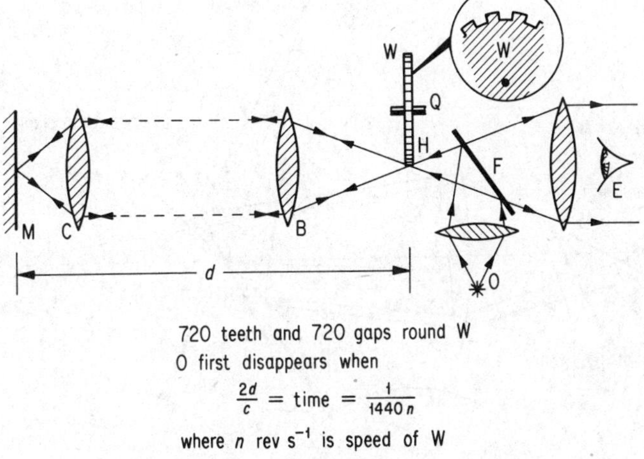

FIG. 6.4. Rotating wheel method for speed of light—Fizeau.

The speed of the wheel is increased from zero. At one speed the light seen through the telescope disappears. With Fizeau's apparatus, this occurred at 12·6 rev s^{-1}. The light has travelled through a gap in the rim of the wheel, and arrives back to find that a neighbouring tooth now occupies the place of the gap. The time taken, t, is thus given by

$$t = \frac{1}{12{\cdot}6 \times 720 \times 2},$$

since the number of gaps and teeth all round the circumference of the wheel is 720×2. (Note the clever way in which a very small time interval was measured fairly accurately.) The distance from the wheel to the plane mirror was 8633 metres, and thus if c is the velocity of light in metres per second,

$$t = \frac{2 \times 8633}{c}$$

$$\therefore \frac{2 \times 8633}{c} = \frac{1}{12{\cdot}6 \times 720 \times 2}$$

$$\therefore c = 3{\cdot}1 \times 10^8 \text{ m s}^{-1}$$

Disadvantage of Fizeau's method. Field of view never becomes perfectly dark, as light is diffusely reflected from teeth. Teeth bevelled to reduce the reflection.

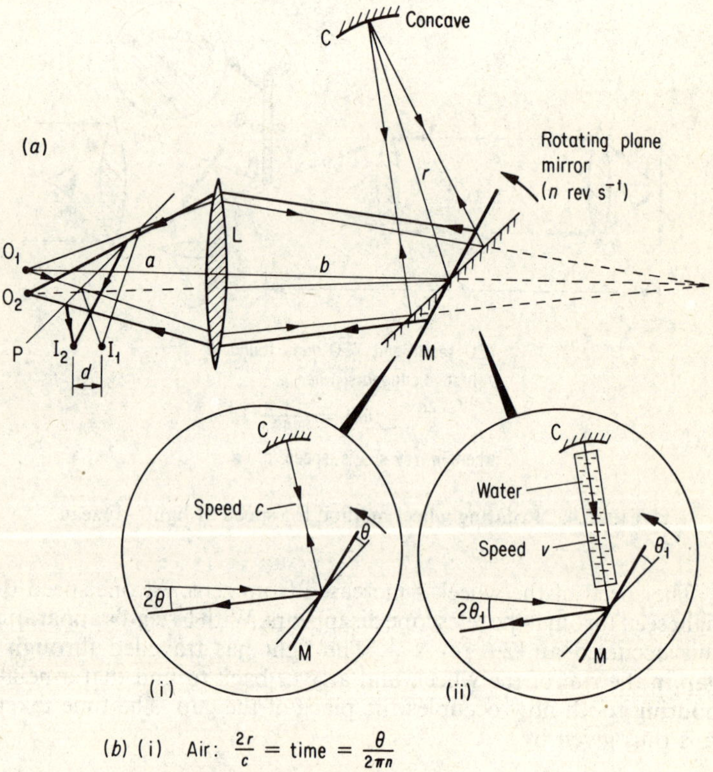

(b) (i) Air: $\frac{2r}{c}$ = time = $\frac{\theta}{2\pi n}$

displacement $d(= O_1O_2) \propto 2\theta$

(ii) Water: displacement greater than in air, so $\theta_1 > \theta$ $\therefore v < c$

FIG. 6.5. Rotating mirror method for speed of light—Foucault.

Foucault's rotating mirror method (1862). Apparatus used:

(1) Powerful source of light, O_1.

(2) Lens, L, providing convergent beam on

(3) Rotating plane mirror M, which reflects the light to

(4) Small concave mirror, C, whose centre of curvature is the axis of rotation of the plane mirror.

(5) A plane glass strip, P, between O_1 and L, which brings the image, I_1, to a convenient place for observation by an eyepiece (Fig. 6.5*a*).

When the mirror is rotating at a fast constant speed, the image is displaced to I_2 by a distance *d*. The velocity of light, *c*, is given by

$$c = \frac{8\pi n r^2 a}{(r+b)d}, \quad \ldots \ldots \quad \text{(i)}$$

where *n* is the number of revolutions per second, *r* is the radius of curvature of the mirror, *a* is the distance from object to lens, and *b* is the distance from lens to rotating mirror. See also Fig. 1.2*b*, p. 2.

The advantage of the method is that, unlike the rotating wheel method, the distances are small, about 20 metres.

The disadvantage is that the image is not very bright.

Velocity of light in water. Pipe of water placed between rotating mirror and curved mirror. Displacement, *d*, observed to be *greater* than in air. Hence, from (i), velocity of light in water was *less* than in air (Fig. 6.5*b*). This has a bearing on Newton's corpuscular theory of light (p. 62).

Michelson's rotating prism method. Apparatus used:

(1) Powerful source of light, O,

(2) Rotating octagonal prism, X, with highly-polished sides, reflecting the light to

(3) Number of plane mirrors, B, L, C, F, M.

(4) Two large concave mirrors, G, H; two inclined plane mirrors, L, C, are in front of one mirror, another plane mirror, F, is at the focus of the second mirror (Fig. 6.6*a*).

Light is incident on one face, A, of the prism, travels some kilometres, and is then observed after reflection at the opposite face, N, of the prism. The speed of the prism is increased from zero until the image is seen by T, and the number of revolutions per second, *m*, is observed. Then time of travel, *t*, of light is given by

$$t = \frac{1}{8m} \quad \text{(Fig. 6.6}b\text{).}$$

(Note the method of measuring a small time interval accurately.) Since the velocity, *c*, $= d/t$, where *d* is the total distance travelled from one face to the opposite face, then

$$c = 8md.$$

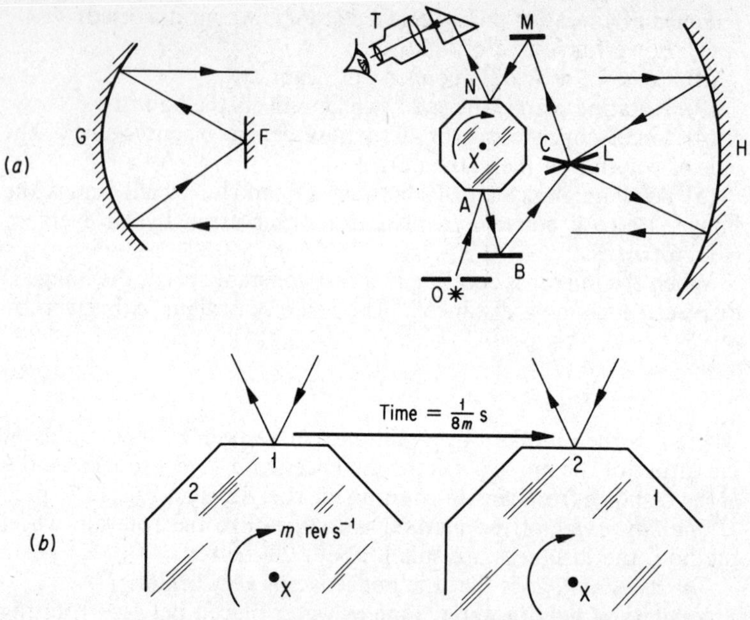

d = light path = A..L..H..G..H..C..N

At m rev s^{-1}, time $= \frac{d}{c} = \frac{1}{8m}$

$\therefore c = 8md$

FIG. 6.6. Rotating prism method for speed of light—Michelson.

Advantage of Michelson's method. It is a null (zero deflection) method, i.e. no displacement is measured, unlike Foucault's method, and m, d were accurately measured.

PHOTOMETRY

Definitions and Units

Luminous flux (F). Luminous energy per second emitted by a source of light. Unit: lumen (lm).

Luminous intensity (I) of a source of light is the luminous flux per unit solid angle emitted in the direction concerned. Unit = candela, symbol "cd" (formerly candle-power (c.p.)).

One candela (cd) is 1/600 000 of the luminous intensity per metre^2 of a black body radiator at the temperature of solidification of platinum.

Lumen. This is the luminous flux emitted in a unit solid angle by a lamp of 1 candela. Symbol: lm.

Illumination (E). This is the luminous flux per unit area incident round the part of the surface concerned. Unit: lux.

1 lux = 1 lm m^{-2} (see later).

Lux. 1 lux is the illumination of a surface due to a lamp of 1 cd placed 1 metre away, when the light is incident normally on the surface. 1 lux = 1 metre-candle.

Transmission factor of a substance is the ratio of the luminous flux passing through it to the luminous flux incident on it.

Reflection factor of a substance is the ratio of the luminous flux per unit area reflected from it to that incident on it.

Luminance (L) of a diffusing surface is the *luminous intensity per unit area projected in the direction concerned.* Thus $L = I/dS \cos\theta$. "*Brightness*" is subjective.

A *uniformly diffusing surface* has a constant brightness at whichever angle it is viewed. Thus $I = I_0 \cos\theta$, where I_0 is a constant.

Lambert's law. The luminous intensity in any direction of an element of a uniformly diffusing surface is directly proportional to the cosine of the angle made with the normal by that direction.

Formulae. Luminous intensity, I, $= F/\omega$

(ω = solid angle = area of spherical cap$/r^2$; round a point, $\omega = 4\pi$. Solid angle unit: steradian (sr)).

$$\text{Illumination, } E, = I\cos\theta/r^2.$$

$$\text{For normal incidence, } E = I/r^2.$$

Note that the formula $E = I\cos\theta/l^2$ is true only for (*i*) a point source of light, (*ii*) a very small area round the point considered, (*iii*) the particular magnitude of I in the direction considered.

Since $F = I\omega$, then when $I = 1$ and $\omega = 1$, $F = 1$. This explains the definition of 1 lumen. Further, as $F = 4\pi$ when $I = 1$ and $\omega = 4\pi$, 4π *lumens are emitted all round a lamp of* 1 *candela*.

Since illumination, E, $= F/A$, when $F = 1$ and $A = 1$ we have $E = 1$. Thus 1 lumen m^{-2} is equivalent to an illumination of 1 lux.

When using the formula for illumination, $E = I\cos\theta/l^2$, note that θ is the angle between *the normal to the surface* and the line joining the lamp to the part of the surface considered.

Lummer-Brodhun photometer. This is an accurate form of photometer. It contains a screen and two right-angle isosceles prisms, which reflect the light towards a "cube". The latter is cut away at the edges on one half. The light from one side of an illuminated screen is reflected from the *outer* part of the cube to the observer's eye. Light from the other side is transmitted through the *inner* part to the observer. When one lamp is moved so that the outer and central parts of the field of view are indistinguishable from each other, a *photometric balance* is obtained.

Comparison of intensities. To compare accurately the luminous intensities I_1, I_2 of two lamps A, B, a third (auxiliary) lamp C is required. Firstly, A and C are placed on opposite sides of the screen and a photometric balance obtained by moving A say. The distance of A from the screen, r, say, is measured. A is then replaced by B. *Keeping* C *the same distance from the screen as before,* B is moved until a balance is again obtained. Suppose its distance is r_2. Then, since the illumination of the screen by A and by B respectively must be equal,

$$\frac{I_1}{r_1^2} = \frac{I_2}{r_2^2},$$

or $I_1/I_2 = r_1^2/r_2^2$.

Measurement of illumination. One form of light-meter consists of a cuprous oxide/copper metal plate connected to a microammeter A. When the oxide side is illuminated electrons are emitted from the surface, and a current flows in A which is proportional to the illumination of the surface. The instrument is calibrated by placing a standard lamp of luminous intensity I at a distance d away; the illumination is then I/d^2. When I is in cd and d in m, then I/d^2 is in lx.

Camera and image brightness. (1) The camera has a variable lens aperture which controls the exposure time and depth of field (p. 42). The *relative aperture* is the ratio d/f, where d is the aperture diameter and f is the focal length. The *f-number* expresses the diameter d as a fraction of the focal length: f-2 (or $f2$ or $f/2$) means $d = f/2$.

(2) Image brightness $\propto$ area of aperture/area of image $\propto d^2/f^2$.

$\therefore$ *Exposure time* $\propto f^2/d^2 \propto$ *squares of f-numbers.*

Thus exposure time for f-2 is one-quarter that required for f-4. On squaring standard f-numbers such as 2, 2·8, 3·5, 4, relative exposure times of 4, 8, 12, 16 are obtained.

WAVES. PHYSICAL OPTICS

7 WAVES AND EFFECTS

Wavelength. Velocity of wave. The wavelength, λ, is the distance between successive crests, or successive troughs, of the wave as it spreads through the medium. If f is the frequency of the wave, then f waves travel past a given point per second. Each occupies a length λ. Thus the *velocity*, v, of the wave (distance travelled per second) is given by

$$v = f\lambda.$$

This relation is true for all waves, whatever their nature.

Examples of waves

Type	Medium	Frequency	Wavelength	Velocity (approx.)
sound	air (0°C)	10^3 Hz	0·34 m	340 m s^{-1}
radio (microwave)	vacuum	10^{10}	0·03	3×10^8
light (yellow)	vacuum	5×10^{14}	6×10^{-7}	3×10^8

Simple harmonic vibration. Each point in a wave undergoes vibration. Simplest vibration is simple harmonic; it has a sinusoidal waveform. *See Book One*. Terms used are:

(*i*) Amplitude, a, = maximum displacement from mean position.
(*ii*) Cycle = one complete (to-and-fro) movement.
(*iii*) Frequency, f, = number of cycles per second. 1 Hz = 1 c/s.
(*iv*) Period, T, = time for one cycle = $\frac{1}{f}$.

For simple harmonic vibration, $y = a \sin \omega t$ or $a \sin \frac{2\pi t}{T}$ or $a \sin 2\pi f t$, where y is the displacement at time t.

Plane-progressive wave. This is a wave whose wave-front is a plane and thus travels in a constant direction, Ox say, through the medium (Fig. 7.1a). The vibrations of points at difference places

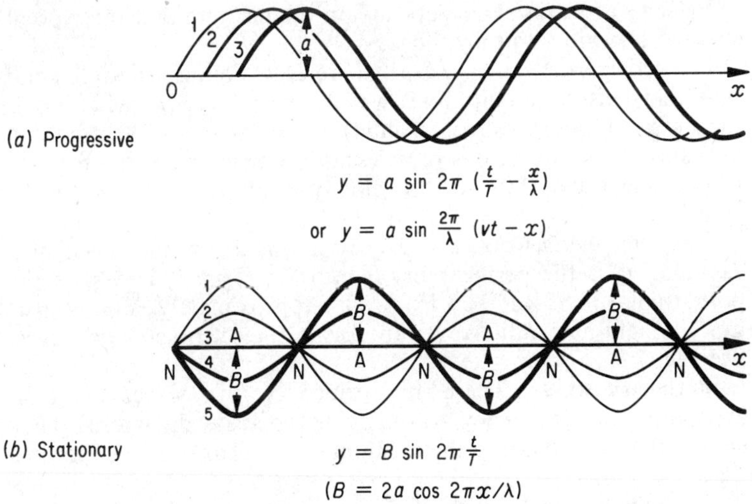

FIG. 7.1. (*a*) Plane-progressive wave (*b*) stationary wave.

are out of phase. At the origin O of the disturbance, suppose the displacement $y = a\sin 2\pi t/T$. At any other point distance x from O, the phase difference δ is given by $\delta = (x/\lambda)\times 2\pi = 2\pi x/\lambda$, since two points λ apart have a phase difference of 2π, and the vibration here *lags* on that at O. Hence the general formula for the displacement at time t of a point distance x from the origin is:

$$y = a\sin 2\pi\left(\frac{t}{T}-\frac{x}{\lambda}\right),$$

or
$$y = a\sin\frac{2\pi}{\lambda}(vt-x),$$

since v = velocity $= f\lambda = \lambda/T$ (Fig. 7.1*a*). (For a wave travelling in the *negative* $(-)\,x$ direction, the phase of the vibration at a point x *leads* that at O by an amount $2\pi x/\lambda$. See p. 56.)

When x is constant, i.e. *at a given place*, $y = a\sin(\omega t-\theta)$, where $\omega = 2\pi v/\lambda$; the displacement y has the same value at a time T later, since $\omega T = 2\pi$. When t is constant, i.e. *at a given time*, $y = a\sin(\alpha-\omega_1 x)$; the displacement y has the same value at a distance λ further, since $\omega_1\lambda = 2\pi$.

EXAMPLE

Suppose $y = 10^{-4}\sin 2\pi(100t-x/3)$, where y, x are in metres and t in seconds.

Then (*i*) a = amplitude $= 10^{-4}$ m, (*ii*) $\lambda = 3$ m, $1/T = f = 100$ Hz, (*iii*) velocity, $v = f\lambda = 100\times 3 = 300$ m s^{-1}, (*iv*) the phase difference between two points 1 m apart $= 2\pi x/\lambda = 2\pi\times 1/3 = 2\pi/3$.

Velocity of waves. This depends on the medium and the type of stress and strain set up by the wave.

Mechanical wave. $v = \sqrt{E/\rho}$, where E is Young's modulus and ρ is the density (longitudinal wave); or $v = \sqrt{K/\rho}$, where K is the bulk modulus (wave in fluid). $K = \gamma p$ for adiabatic bulk modulus of gas, where p is the pressure. A sound wave is a form of mechanical wave. At 0 °C, its velocity in air is about 331 m s^{-1} (Fig. 7.2*c*).

Electromagnetic wave. $c = 1/\sqrt{\mu_0\varepsilon_0}$, where c is the speed in a vacuum, μ_0 is the permeability ($4\pi\times 10^{-7}$ H m^{-1}) and ε_0 is the permittivity ($8{\cdot}85\times 10^{-12}$ F m^{-1}). Approximately, $c = 3\times 10^8$ m s^{-1}. Light and radio waves are forms of electromagnetic waves. See Fig. 7.2*a*, *b*.

Stationary waves. These are formed by two waves of equal amplitude and frequency travelling in opposite directions. They stay at the same place in the medium. Their characteristic features are:

(1) Some points, N, in the medium are permanently at rest; these are called "nodes" (See Fig. 7.1*b*).

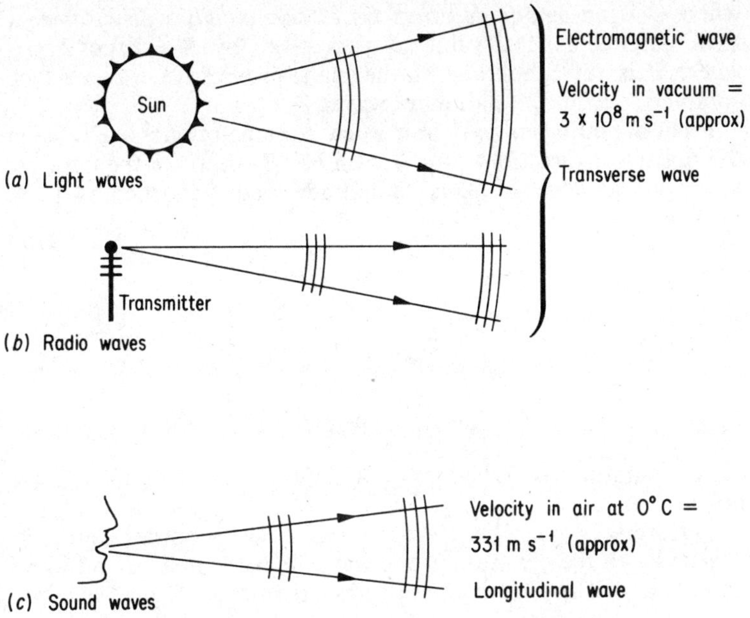

FIG. 7.2. Examples of waves.

(2) Points between successive nodes, N, N, are vibrating with the same frequency and phase; the amplitude of vibration increases from a node to the mid-point A of successive nodes N, N, where it is a maximum. A is called an "antinode".

(3) The distance between successive nodes, NN, or successive antinodes, AA, $= \frac{\lambda}{2}$. The distance NA from a node to the next antinode $= \frac{\lambda}{4}$.

In contrast, a plane-progressive wave is one in which the amplitude is the same at all points in the medium, and neighbouring points vibrate out of phase. There is no stationary pattern as in the case of the stationary wave. See Fig. 7.1.

The displacement curve of a stationary wave is the variation of the displacements of the particles at a given time. The curve is a series of stationary loops, as zero displacement occurs at the nodes.

Stationary wave displacement. This displacement y can be represented by

$$y = B \sin \frac{2\pi t}{T},$$

where B is an *amplitude which varies sinusoidally with distance x from the source*. The value of B is given by $B = 2a\cos 2\pi x/\lambda$, where B is the amplitude of the plane-progressive waves which combine to form the stationary wave.

Proof of stationary wave properties. A plane-progressive wave in the positive-x direction is $y_1 = a\sin 2\pi(t/T - x/\lambda)$; in the opposite direction, a wave of equal amplitude and frequency is $y_2 = a\sin 2\pi(t/T + x/\lambda)$.

$$\therefore \text{ resultant, } y = a\left[\sin 2\pi\left(\frac{t}{T}-\frac{x}{\lambda}\right)+\sin 2\pi\left(\frac{t}{T}+\frac{x}{\lambda}\right)\right]$$

$$= 2a\sin\frac{2\pi t}{T}.\cos\frac{2\pi x}{\lambda} = B\sin\frac{2\pi t}{T},$$

where $B = 2a\cos\frac{2\pi x}{\lambda}$. When $x = 0, \lambda/2, \lambda, \ldots$, B is a maximum; these are antinodes. When $x = \lambda/4, 3\lambda/4, \ldots$, B is zero; these are nodes.

Transverse wave. This is one in which the vibrations are *perpendicular* to the direction of the wave. Waves spreading on water from a stone dropped in it are transverse waves. Light waves are transverse waves (p. 75) (Fig. 7.3*a*).

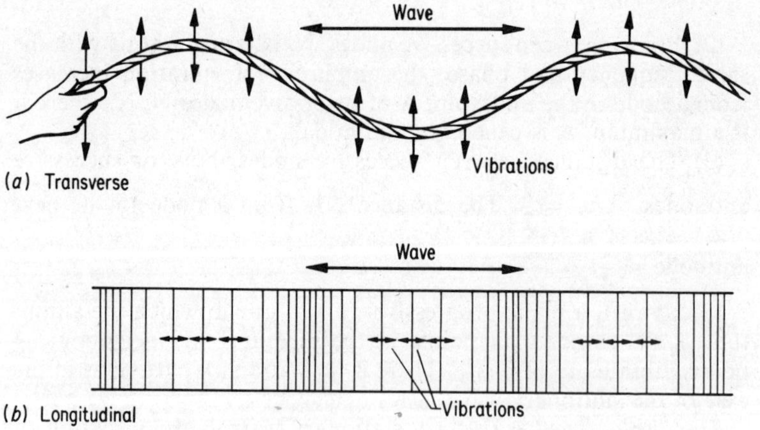

FIG. 7.3. Transverse and longitudinal waves.

Longitudinal wave. This is one in which the vibrations of the particles are in the *same* direction as the wave. Sound waves are longitudinal waves (Fig. 7.3*b*).

Progressive and stationary waves may be longitudinal or transverse.

Diffraction of waves. Waves which meet an opening may pass through with negligible change in direction, i.e. negligible diffraction. This is the case when the width of the opening is large compared with the wavelength λ of the waves. Light, for example, has a wavelength of the order 6×10^{-4} mm. It therefore passes through normal openings without change of direction, i.e. "rectilinear propagation" of light occurs, and sharp shadows are thus produced (Fig. 7.4*a*).

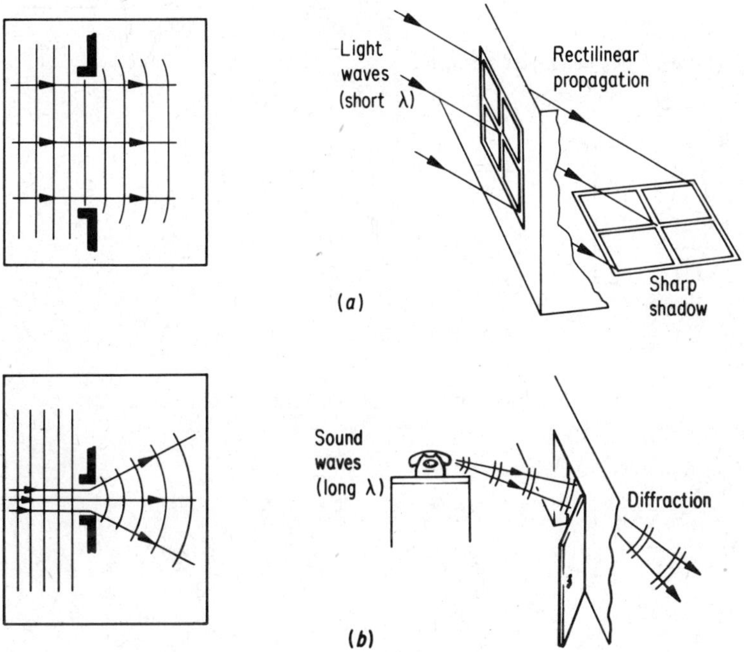

FIG. 7.4. Wavelength and diffraction.

On the other hand, sound waves have wavelengths of the order of 1 metre. Hence appreciable diffraction of sound waves occurs, i.e. sound may be heard round corners (Fig. 7.4*b*).

Electromagnetic waves. By using a microwave (3 cm) emitter and receiver, a metal plate as a reflector, a large hollow prism filled with liquid as a refractor, "openings" as coherent sources for interference and diffraction experiments, and aluminium grilles for polarization experiments, it can be demonstrated that electromagnetic waves can undergo reflection, refraction, interference, diffraction and polarization. See Fig. 7.5–7.8. Light waves are electromagnetic waves of extremely short wavelength, about 6×10^{-7} m.

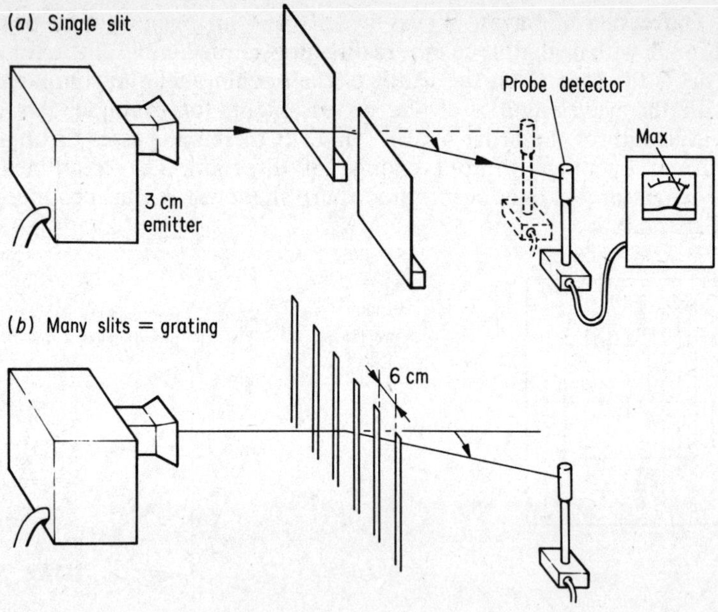

FIG. 7.5. Electromagnetic (3 cm) waves—diffraction.

$S_2P - S_1P = \lambda$ (Constructive interference)

$S_2Q - S_1Q = 2\lambda$

$S_2R - S_1R = \lambda$

FIG. 7.6. Interference with two parallel slits (Young).

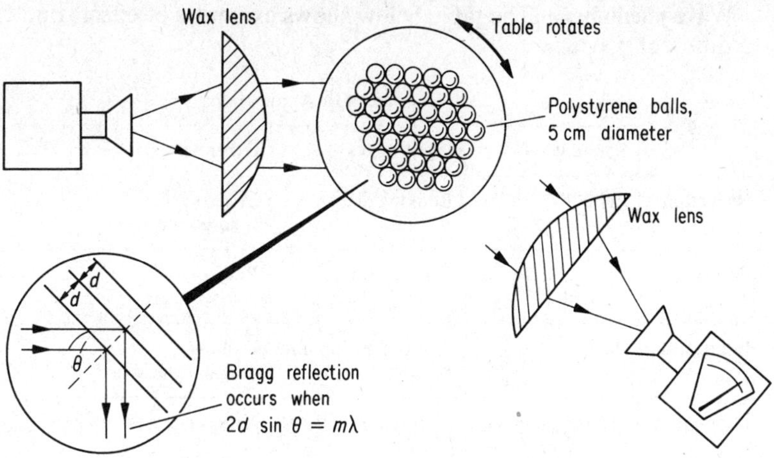

FIG. 7.7. Diffraction by model 'crystal'—Bragg reflection.

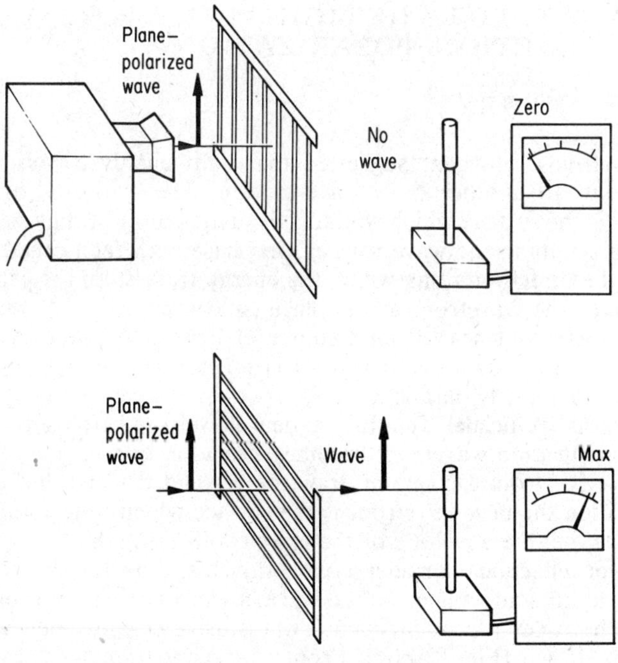

FIG. 7.8. Demonstration of plane polarization.

Wave phenomena. The table below shows examples of effects due to different waves:

	Phenomena or Applications		
	Sound waves	Light waves	Other waves
Reflection	Echo	Looking glass	Microwaves—speed trap
Refraction	Travel further at night	Lens action	Radio—ionosphere action
Interference	Beats	Colours in thin films	Radio—signal weakened
Diffraction	Heard round corners	Diffraction grating	X-rays—crystal analysis
Polarization	None	Polaroid filter	Radio—aerial direction finding

8. WAVE THEORY OF LIGHT. INTERFERENCE. DIFFRACTION. POLARIZATION

WAVE THEORY

Wavefronts. Huygens suggested that light energy travels from one point to another by a wave-motion. The *wavefront* at any instant is the surface which contains the light energy at that instant, and all points in the wavefront are in phase with each other. The *ray* is the direction along which the energy travels, and is perpendicular to the wavefront at the place concerned.

The wavefront near a point source of light is a sphere. A long way from the source the wavefront is substantially plane, and the rays are then fairly parallel.

Huygens' principle. To draw a new wavefront, Huygens said: Every point on a wavefront becomes a new or *secondary centre of disturbance*; from each point draw the wavelet reached in a given time. Then the new wavefront is the surface which touches all the wavelets, i.e. the envelope of the wavelets. See Fig. 8.1*a*.

Law of reflection. Consider a plane wavefront incident on a plane mirror at an acute angle. As the wave advances, each point on the mirror becomes a secondary centre of disturbance, and a new wavefront is drawn (Fig. 8.1*b*(i)). From congruent triangles it can be shown that the angle of incidence = the angle of reflection.

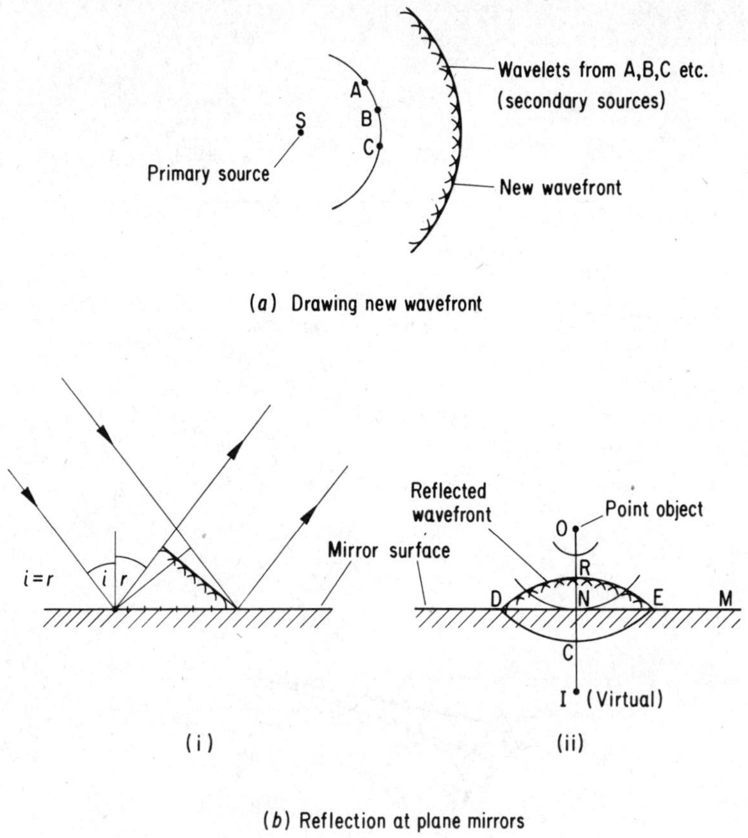

FIG. 8.1. (*a*) Huygens' principle (*b*) reflection at plane surfaces.

Point object and image in mirror. A point object O emits a spherical wavefront which reaches N at a mirror M at some instant (Fig. 8.1*b*(ii)). After this, points on the mirror emit wavelets towards O. In the absence of the mirror, the wavefront would reach a position C. From Huygens' principle, the reflected wavefront is spherical and reaches R, however, where NR = NC. The wavefront DRE appears to have its centre at I, behind the mirror, and from symmetry, ON = NI.

Law of refraction. Consider a plane wavefront incident on an air-water plane boundary. As the wave advances, each point on the water becomes a secondary centre of disturbance. Wavelets are hence drawn in the water with a radius vt, where v is the velocity of light *in water* and t is the time concerned. The wavefront veers round towards the normal (Fig. 8.2*a*).

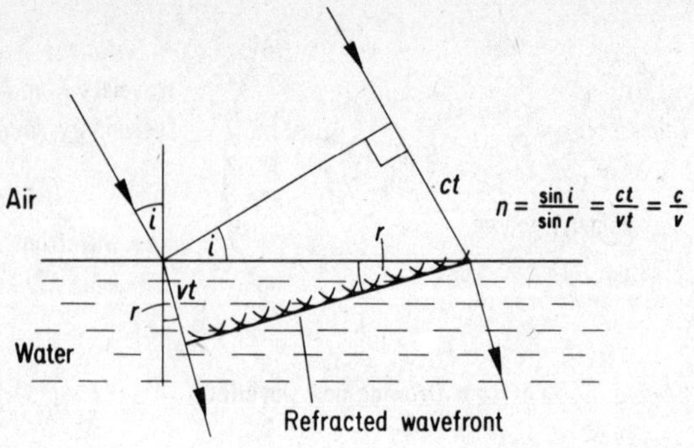

(*a*) Air – water

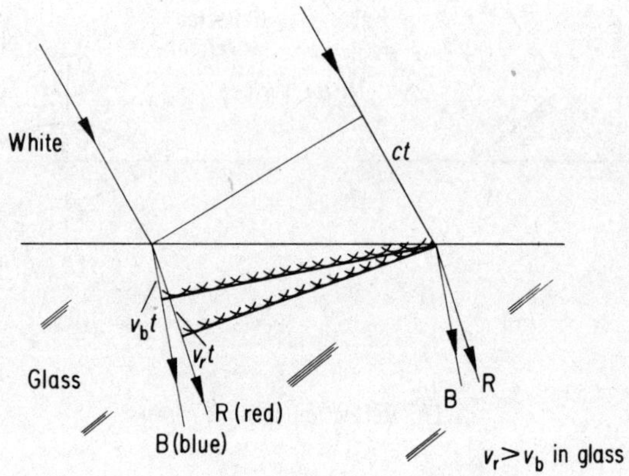

(*b*) Dispersion

FIG. 8.2. Refraction and dispersion.

By right-angled triangles, it can be shown that

$$\frac{\sin i}{\sin r} = \frac{c}{v},$$

where c is the velocity of light in air. But c, v are constants. Hence $\sin i/\sin r$ = constant.

From the wave-theory, the refractive index, $n = c/v$. But n is known to be greater than 1; hence c is greater than v. Now Newton had proposed a corpuscular or particle theory of light, which led to the result $n = v/c$. On Newton's theory, then, v was greater than c. Foucault observed in his rotating mirror experiment that the velocity of light in air was greater than in water (see p. 49). Consequently Newton's theory was wrong.

Dispersion. This is due to the different speeds of colours in glass and other media. Blue light has a smaller speed than red light. Huygens' construction for the new wavefronts in glass from air shows that the wavefront for blue light veers towards the normal more than that for red light, that is, the angle of refraction for blue light is less than that for red light. See Fig. 8.2*b*.

Doppler's principle in light. When a source of monochromatic light of wavelength λ is receding from an observer on the earth with a velocity v, the apparent wavelength λ' is given by

$$\lambda' = \frac{c+v}{c}.\lambda \quad . \quad . \quad . \quad . \quad . \quad \text{(i)}$$

or

$$\text{wavelength shift} = \lambda' - \lambda = \frac{v}{c}.\lambda \quad . \quad . \quad . \quad . \quad \text{(ii)}$$

Since λ' is greater than λ, the wavelength shifts towards the red. When the light moves directly towards the observer, the apparent wavelength λ'' is given by

$$\lambda'' = \frac{c-v}{c}.\lambda \quad . \quad . \quad . \quad . \quad . \quad \text{(iii)}$$

or

$$\text{wavelength shift} = \lambda - \lambda'' = \frac{v}{c}.\lambda \quad . \quad . \quad . \quad \text{(iv)}$$

This is a blue shift.

The Fraunhofer lines in the solar spectrum (p. 32) differ very slightly when the light is observed from opposite edges of the sun's diameter. This is due to the rotation of the sun about an axis through its centre—one edge sends light towards, and the other away, from the observer, so that a different Doppler effect is observed.

Photoelectric effect. In the photoelectric effect, light behaves as if it is due to *particles* (photons) of energy $h\nu$, where ν = frequency $= c/\lambda$. In this case, electrons may be emitted from metal surfaces whose maximum velocity v_m is given by $h\nu - w_0 = \frac{1}{2}mv_m^2$, where w_0 is the "work function" of the metal. See *Book One: Mechanics, Electricity, Atomic Physics,* for discussion of the photoelectric effect.

INTERFERENCE OF LIGHT

Conditions for interference of light. Two sources of light will produce interference bands if (*i*) they are *coherent sources*, i.e. they have the same frequency and amplitude and are always in phase with each other, (*ii*) their distance apart is very small and of the same order as the wavelength of light. This means in practice that each source must be an image of a common source *or* one must be the image of the other.

Suppose P, Q are two coherent sources and S is a point nearly equidistant from them. Then (SP − SQ) is the path difference between the two waves from P, Q which arrive at S.

If optical path difference = $m\lambda$, a *bright* band is observed at S.

If optical path difference = $(m+\frac{1}{2})\lambda$, a *dark* band is observed. (Here m = 0, 1, 2, etc.; thus an odd number of half-wavelengths implies a dark band.) Optical path = nd if d is geometrical path.

Sometimes, as in Newton's rings, a wave is reflected at a *denser* medium before reaching S; this produces a phase change of 180° in the wave, which is equivalent to an extra distance of $\lambda/2$.

Young's experiment. *Demonstrated* by light from a line filament of a lamp A concentrated on to a narrow slit S of a collimator tube in front of two narrow slits S_1, S_2 (Fig. 8.3*a*). Interference

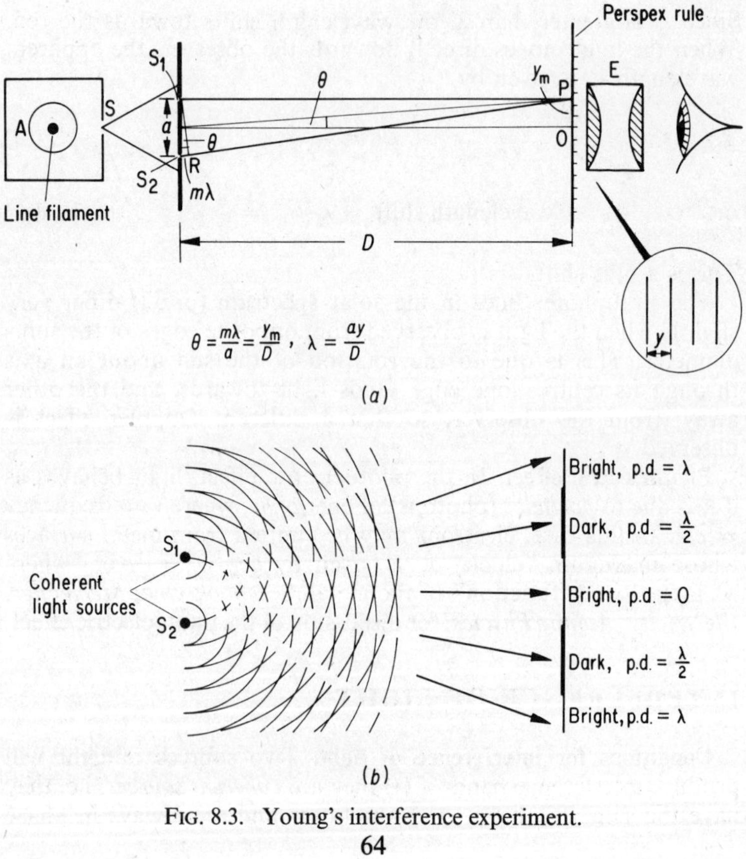

FIG. 8.3. Young's interference experiment.

bands are seen through an eyepiece E when S_1, S_2 are of the order of a millimetre apart or less. Their separation can be measured roughly by a transparent ruler. They are due to waves from S_1, S_2 overlapping, as shown in Fig. 8.3*b*. Red and blue bands are obtained by using colour filters in front of S.

Interference bands. The separation of the bright bands can be found from Fig. 8.3*a*. Suppose y_m is the distance of the *m*th bright band from O. If $PS_1 = PR$, then path difference $= S_2R$. Now

$$\sin\theta = \frac{S_2R}{a} = \frac{m\lambda}{a}, \qquad \tan\theta = \frac{y_m}{D}.$$

Also, $\sin\theta = \tan\theta$ when θ is very small. Thus $y_m = m\lambda D/a$.

$$\therefore \text{ band separation, } y, = y_m - y_{m-1} = \frac{m\lambda D}{a} - \frac{(m-1)\lambda D}{a}$$

$$\therefore y = \frac{\lambda D}{a} \quad . \quad . \quad . \quad . \quad . \quad . \quad . \quad . \quad . \quad . \quad . \quad \text{(i)}$$

and

$$\lambda = \frac{ay}{D} \quad . \quad . \quad . \quad . \quad . \quad . \quad . \quad . \quad . \quad . \quad . \quad \text{(ii)}$$

Special points. (1) Using filters, the separation *y* of red bands is greater than that for blue bands, and hence the wavelength of red light is greater than that of blue light.

(2) If S is moved nearer S_1, S_2, the separation of the bands is unaffected, from (i), but the brightness is increased.

(3) If the distance S_1S_2 is diminished, or *D* is increased, the separation of the bands increases, from (i).

(4) If the slit S is opened wider, it becomes equivalent to a lot of narrow slits each giving rise to its own interference pattern, so that overlapping of bright and dark bands occur and at one stage the bands cannot be seen. The condition for bands to be visible is:

$$\textit{slit width} < \frac{\lambda D'}{a},$$

where D' is the distance from S to the slits S_1, S_2.

Air-wedge film. *Apparatus required*: (*i*) Two microscope slides in contact at one edge, and separated at the other end by thin foil; (*ii*) sodium burner and plane glass to reflect the light downward; (*iii*) travelling microscope focussed on the top of the lower slide (Fig. 8.4*a*). *Interference pattern*: Bright and dark straight bands, parallel to the edge of the air-wedge formed. The bands are contours of paths of equal thickness.

Coherent sources. (*i*) Light reflected upward at the lower edge T of the top slide, (*ii*) light travelling down from T through the air-film and reflected upward from the bottom side. See Fig. 8.4*a*.

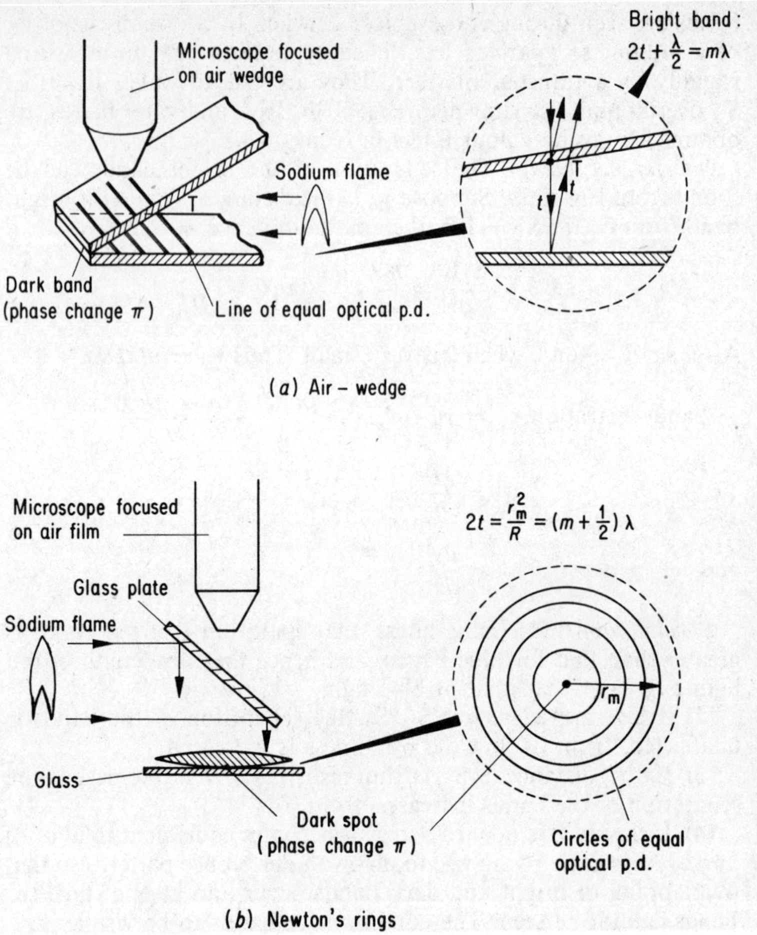

FIG. 8.4. (*a*) Air-wedge film (*b*) Newton's rings.

Theory. Path difference $= 2t+\dfrac{\lambda}{2}$ (reflection at a denser medium).

$$\therefore\ 2t+\frac{\lambda}{2} = m\lambda, \text{ for bright band, or } 2t = (m-\tfrac{1}{2})\lambda,$$

and $$2t+\frac{\lambda}{2} = (m+\tfrac{1}{2})\lambda, \text{ for dark band, or } 2t = m\lambda.$$

Applications. (1) If there are m bright bands in the wedge, the thickness t of the thin foil is given by $2t = (m-\frac{1}{2})\lambda$, or $t = (m-\frac{1}{2})\dfrac{\lambda}{2}$.

(2) The angle θ of the wedge can be found by measuring the length l occupied by the m bands with a travelling microscope. Then $\theta = t/l$ radians and t is obtained from m, λ.

(3) Measurements of small changes of length, as in experiments to measure the coefficient of linear expansion of a solid, are made very accurately by means of an air wedge of small angle. If x bright bands cross a given point in the air wedge when expansion occurs, the base of the air wedge being the end of the solid, increase in length $= x\lambda/2$.

EXAMPLE

Two optically flat plates, inclined at a small angle of 10′ to each other, are viewed normally in reflected light of wavelength 6×10^{-7} m. Calculate the number of bright bands per metre.

With the usual notation, $2t = (m-\frac{1}{2})\lambda$ is the condition for bright bands. Thus the path difference t_1 for successive bands is given by $2t_1 = \lambda$.

But $$\frac{t_1}{x} = \text{angle of wedge in radians,}$$

where x is the distance between successive bands.

$$\therefore \frac{t_1}{x} = \frac{10}{60}\times\frac{\pi}{180}$$

$$\therefore t_1 = \frac{\pi x}{6\times180}.$$

But $2t_1 = \lambda$, $\therefore \dfrac{2\pi x}{6\times180} = \dfrac{6}{10^7}$

$$\therefore \frac{1}{x} = \frac{2\pi\times10^7}{6\times6\times180} = 9700 \text{ (approx.)}$$

$$\therefore \text{bands per metre} = \frac{1}{x} = 9700.$$

Newton's rings. *Apparatus required*: (*i*) Converging lens with surface of large radius of curvature placed on plane glass surface; (*ii*) sodium burner and plane glass to reflect the light downward; (*iii*) travelling microscope focused on the plane glass surface below the lens (Fig. 8.4*b*). *Interference pattern*: Bright and dark rings; the point of contact between the lens and glass surface is a dark spot.

Coherent sources. (*i*) Light reflected upwards from the lower surface S of the lens, (*ii*) light travelling down from S through the airfilm and reflected upward from the plane glass.

Theory. Path difference $= 2t+\dfrac{\lambda}{2}$ (reflection at a denser medium).

$$\therefore 2t+\frac{\lambda}{2} = m\lambda \text{ for bright ring,}$$

or $$2t = (m-\tfrac{1}{2})\lambda.$$

But $$2t = \frac{r^2}{a},$$

to a good approximation, where r is the radius of the corresponding ring and a is the radius of curvature of the *lower* lens surface.

$$\therefore\ r^2 = (m-\tfrac{1}{2})a\lambda \quad . \quad . \quad . \quad . \quad . \quad . \quad . \quad . \quad \text{(i)}$$

Applications. (1) The radius r_1 of the first bright ring is thus given by $r_1{}^2 = \frac{1}{2}a\lambda$. Measurement of a and the radius r of the fifteenth ring, say, enables λ to be calculated.

(2) If the lower lens surface is not perfectly spherical, the rings are not perfectly circular. This is used for testing lens' surfaces.

(3) If blue light replaces red light the rings become narrower. From (i) this proves that blue light has a smaller wavelength than red light.

(4) If the lens and plane surface are separated slowly, the rings appear to shrink towards the central spot, which becomes alternately bright and dark.

Fresnel's biprism. A prism of *very* large angle, with a narrow slit in front of the refracting edge, provides two virtual images close together by refraction in each half of the prism. These are coherent sources. Straight interference bands, equally spaced, are produced in front of the prism, and may be observed by an eyepiece about a metre away. The wavelength can be measured from the relation

$$\lambda = \frac{ay}{D},$$

where y is the distance between successive bright bands, a is the distance between the virtual images, and D is the distance from the telescope to the images. (See p. 65 for proof.) The distance a can be measured by moving a convex lens between the biprism and eyepiece so that the distance apart a of the virtual images is magnified to a distance d, which is measured. If the lens produces a magnification m (equal to v/u), then $a = d/m$.

If the slit is a distance t in front of the biprism, and the latter has a narrow refracting angle α equal to $(180° - A)/2$, where A is the large biprism angle, then a = distance apart of virtual images $= 2(n-1)\alpha t$.

Lloyd's mirror. Lloyd used a source of light S close to a plane mirror so that the light was nearly at grazing incidence. S and its image I in the mirror were thus two close coherent sources, and interference bands were obtained on a screen beyond the mirror.

Thin films. In thin parallel-sided film of refractive index n,

$$\text{path difference} = 2nt\cos r.$$

With reflection at a denser medium taken into account,

$$2nt\cos r = (m-\tfrac{1}{2})\lambda \text{ for bright band,}$$

$$2nt\cos r = m\lambda \text{ for dark band.}$$

With an extended source, interference occurs between (1) the light beam reflected from the upper surface of the film and (2) the remainder of the light which travels into the film and is reflected from its lower surface, when the two beams are combined by an eyelens, for example. A *thin* film is essential to obtain a very small path difference. The bands are located at the film, and are contours of paths of equal thickness if r is constant.

Colours in thin films such as oil-films or soap-bubbles are due to interference. If light enters the eye pupil from a point on the film where the path difference of the two interfering beams is $2nt\cos r = (m-\frac{1}{2})\lambda$, then light of the particular wavelength λ is seen. The thin film is illuminated by white light from an extended object such as the sky, and thus λ may correspond to a colour such as blue. The blue band observed is the contour of paths of equal thickness t in the film, as r is fairly constant. If the thickness is such that $2nt\cos r = m\lambda$, the wavelength λ is missing from the white light, and the complementary colour is observed.

EXAMPLE

A soap film has a refractive index of 1·4 and is 4×10^{-7} m thick. If it is viewed from an angle of 60° to the normal, find the wavelengths in the visible spectrum ($4\times10^{-7}-7{\cdot}5\times10^{-7}$ m) absent from the reflected light, and state the observed colour of the film.

The path difference of the light reaching the eye $= 2nt\cos r$, where r is the angle of refraction. The wavelength, λ, missing is hence given by

$$2nt\cos r+\frac{\lambda}{2} = (m+\tfrac{1}{2})\lambda,$$

allowing for a phase change of 180° by reflection.

$$\therefore\ 2nt\cos r = m\lambda \quad . \; . \; . \; . \; . \; . \; . \; . \quad \text{(i)}$$

When $m = 1$,

$$\lambda = 2nt\cos r \quad . \; . \; . \; . \; . \; . \; . \quad \text{(ii)}$$

The angle of refraction, r, is given by

$$\frac{\sin 60^\circ}{\sin r} = 1{\cdot}4,$$

from which $r = 38^\circ\,12'$.

From (ii), $\lambda = 4\times1{\cdot}4\times4\times10^{-7}\cos 38^\circ\,12'$ m

$= 8{\cdot}8\times10^{-7}$ m . . . . (iii)

When $m = 2$, the wavelength missing is given from (i) by

$$\lambda_1 = \frac{2nt\cos r}{2} = 4{\cdot}4\times10^{-7} \text{ m from (iii).}$$

When $m = 3$, the wavelength missing is given by

$$\lambda_2 = \frac{2nt\cos r}{3} = 2{\cdot}9\times10^{-7}\text{ m from (iii).}$$

Thus the wavelength $4{\cdot}4\times10^{-7}$ m in the visible spectrum is absent from the reflected light, and as this is blue the film appears greenish-yellow.

DIFFRACTION OF LIGHT

Diffraction may be defined as the interference of light from sources *on the same wavefront*, which are coherent.

Diffraction at a wide slit. Rectilinear propagation of light occurs. See Fig. 7.4*a*. The absence of light round the edges of the slit is due to interference by coherent sources on a very wide wavefront, very much wider than the wavelength of light.

Diffraction at a narrow slit. In addition to a central bright image, a number of separated diffraction images are obtained in the geometrical shadow of the slit. This is due to interference between the coherent sources on a *narrow* wavefront, comparable with the wavelength of light.

The intensity or brightness of the central image diminishes from its centre outwards. See Fig. 8.5*a*. The angular width (spread) of the central image is 2θ, where $\theta = \lambda/a$ and a is the slit width. The middle of the other diffraction images correspond to larger angles of θ, $3\lambda/2a$, $5\lambda/2a$, ... The intensity of the first diffraction image is very small compared with that of the central image.

With light waves, λ is very small and about 6×10^{-7} m. From $\theta = \lambda/a$, it follows that θ is very small when a is large compared with 6×10^{-7} m (wide slit), and very large when a is very narrow and comparable with λ (narrow slit). In the latter case, light spreads round corners or openings. With sound waves, λ is long, for example, 0·3 m. Thus even with wide openings (a large), sound waves spread round corners.

Diffraction at many slits. As the number of slits increases, 2, 3, 4, . . . 15, . . ., the diffraction pattern becomes a series of bright images (principal maxima) increasing in intensity, separated by a number of less bright images (secondary maxima) which diminish in intensity. See Fig. 8.5*b*. The sharpness of the principal maxima images increases at the same time. For many thousands of lines per cm, as in the diffraction grating, the only images seen are those of the principal maxima, which are very sharp and separated considerably from each other. See Fig. 8.6.

Diffraction grating. This is a series of narrow openings, produced by ruling fine parallel lines, perhaps 1000 lines to the millimetre, close to each other on a transparent surface such as glass. The line separation or spacing $d = a+b$, where a is the width of the

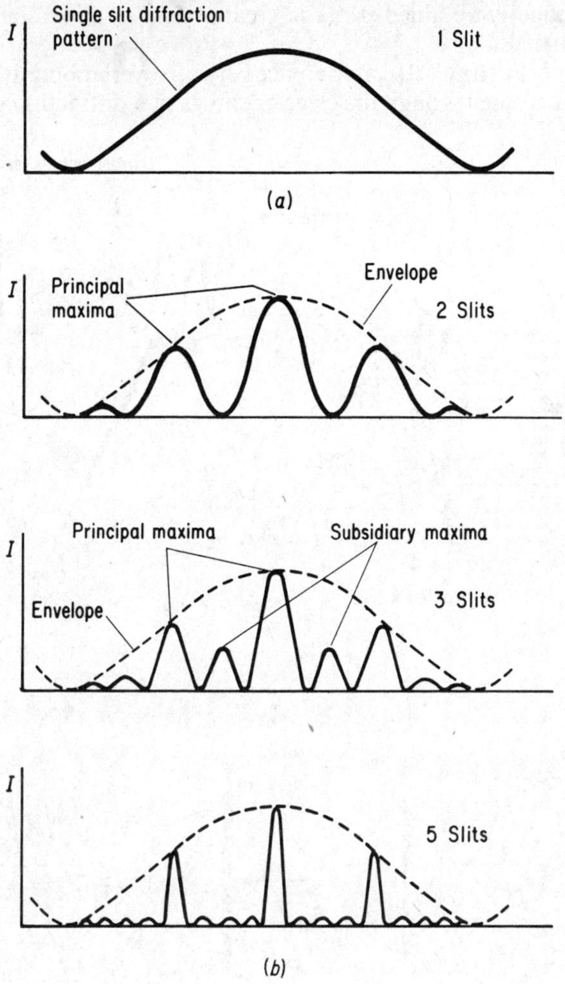

FIG. 8.5. Diffraction at slits.

(clear) opening and b is that of the (opaque) line, $=$ 1/(number of lines per mm) (Fig. 8.6*a*). By considering a plane wavefront of monochromatic light on the grating, and a direction inclined at θ to the normal, we find the path difference is $d \sin\theta$ for *all* sources in adjacent openings separated by a distance d.

$$\therefore\ d\sin\theta = m\lambda \text{ gives a bright image.}$$

First diffraction image when $\sin\theta = \lambda/d$ ($m = 1$). See Fig. 8.6*a*.
Second diffraction image when $\sin\theta_1 = 2\lambda/d$ ($m = 2$).

No image is obtained if $m\lambda$ is greater than d, since $\sin\theta$ must be less than 1.

Fig. 8.6*b* illustrates the interference between overlapping diffracted beams emerging through the clear openings of a diffraction grating.

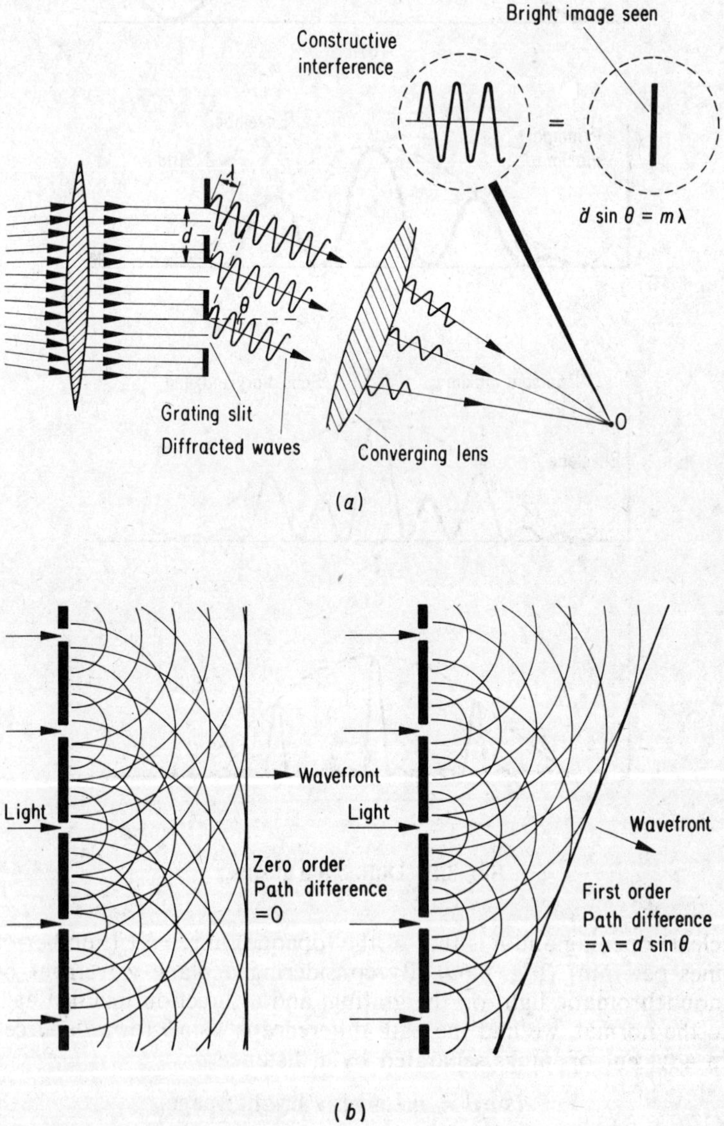

FIG. 8.6. Diffraction grating action.

EXAMPLE

Sodium light, with wavelengths 5·890 and $5{\cdot}896\times10^{-7}$ m, falls normally on a diffraction grating. It is observed that in the first-order diffracted beam, the two sodium lines are separated by exactly 2 minutes of arc. What is the spacing of the grating?

The first-order beam is given by

$$d\sin\theta = \lambda \quad . \quad . \quad . \quad . \quad . \quad . \quad . \quad . \quad \text{(i)}$$

By differentiation, $d\cos\theta.\delta\lambda = \delta\lambda$.

$$\therefore\ d\cos\theta = \frac{\delta\lambda}{\delta\theta}.$$

Now $\delta\lambda = 5{\cdot}896\times10^{-7} - 5{\cdot}890\times10^{-7} = 0{\cdot}006\times10^{-7}$ m,

and $\delta\theta = 2 \text{ minutes} = \frac{1}{30}\text{ degree} = \frac{\pi}{30\times180}$ radians.

$$\therefore\ d\cos\theta = \frac{0{\cdot}006\times10^{-7}\times30\times180}{\pi} \quad . \quad . \quad . \quad . \quad . \quad \text{(ii)}$$

Dividing (i) by (ii), and using $\lambda = 5{\cdot}890\times10^{-7}$ m,

$$\frac{\sin\theta}{\cos\theta} = \tan\theta = \frac{5{\cdot}890\times10^{-7}\times\pi}{0{\cdot}006\times10^{-7}\times30\times180}$$

$$\therefore\ \theta = 29^\circ\,43'$$

$$\therefore\ d = \frac{\lambda}{\sin\theta} = \frac{5{\cdot}890\times10^{-7}}{\sin 29^\circ\,43'} = 1{\cdot}2\times10^{-6}\text{ m.}$$

Measurement of wavelength. Diffraction grating and spectrometer used. *Adjustment.* (1) Place the telescope at 90° to the collimator. (2) Place the grating on the spectrometer table perpendicular to two screws A, B, turn the table until the image of the collimator slit is seen by reflection in the grating, and adjust A or B to bring the image central in the field of view. (3) Now turn the grating perpendicular to the collimator, obtain the first diffraction image in the telescope, and move the third screw until the image is central—the lines on the grating are then vertical.

Measurement. The first diffraction image position on either side of the normal is measured, and the angle halved to obtain θ. Second diffraction image also obtained. Then $\lambda = d\sin\theta$ or $\frac{1}{2}d\sin\theta_1$, where θ, θ_1 correspond to the first and second diffraction images respectively.

Wavelengths may easily be compared, since $\lambda:\lambda' = \sin\theta:\sin\theta'$ for a given order of diffraction image.

Resolving power of telescope. A lens cannot resolve, or distinguish separately, the images of two distant objects which subtend an angle at the objective less than $1{\cdot}22\lambda/D$, where λ is the wavelength and D is the objective diameter (Fig. 8.7*a*). The *resolving power* of the telescope is defined as $1{\cdot}22\lambda/D$. An objective of large diameter increases the resolving power. The Palomar *mirror* telescope has a high resolving power also given by $1{\cdot}22\lambda/D$, where D is the aperture of the mirror used as the objective.

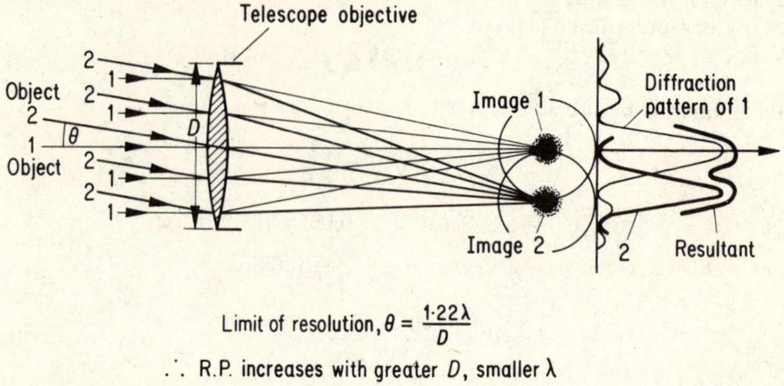

(*a*)

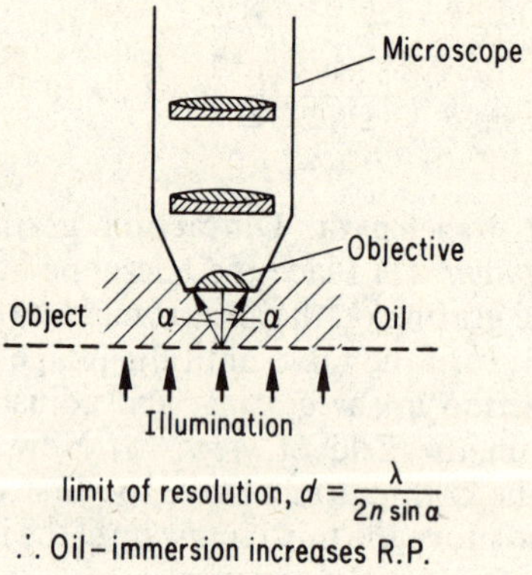

(*b*)

FIG. 8.7. Resolving power of (*a*) telescope (*b*) microscope.

Resolving power of microscope. This is $\lambda/(2n \sin \alpha)$, where λ is the wavelength, n is the refractive index of the medium in which the objects are situated, and 2α is the angle subtended by the objective at the objects (Fig. 8.7*b*). In the Abbe *oil-immersion objective*, the objective dips into cedar-oil of the same refractive index, about 1·5, and the object can then be brought much nearer to the objective, thus increasing α and hence the resolving power.

Resolving power of diffraction grating. The resolving power = $\frac{\lambda}{d\lambda} = Nm$, where N is the total number of lines on the grating and m is the order of the spectrum. For a given order of spectrum, the resolving power depends on the product of the width of the grating and the number of lines per metre.

POLARIZATION OF LIGHT

Polarization by Polaroids. If light is viewed through two parallel Polaroids A, B, it appears to be only slightly affected. If A is kept fixed, however, and B is rotated about the incident light as axis, darkness is observed on looking at B when the Polaroids are perpendicular to each other (Fig. 8.8*a*).

We conclude that *light waves are transverse waves,* that is, the vibrations of light are in a plane perpendicular to the direction of

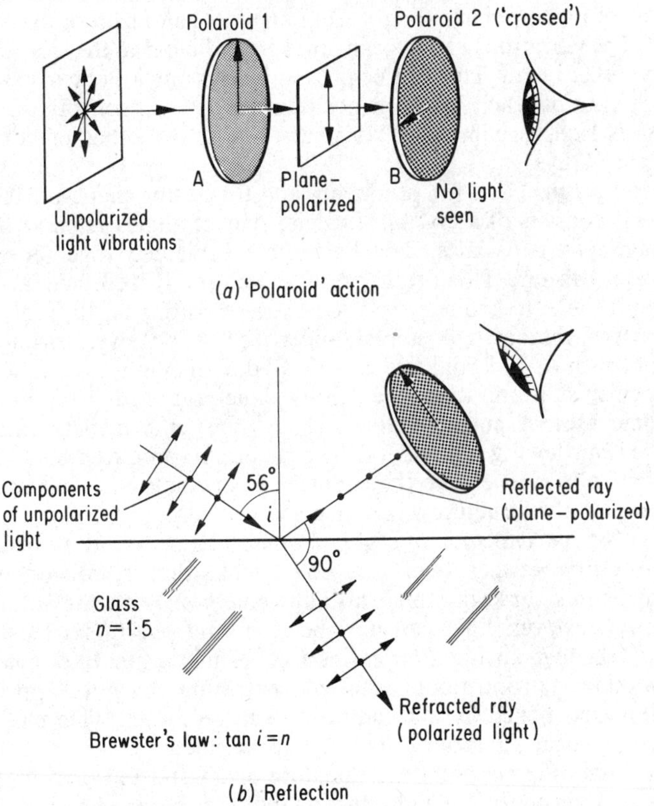

FIG. 8.8. Polarization by (*a*) *Polaroid* (*b*) reflection.

the light. If light consisted of longitudinal waves, that is, the vibrations were parallel to the direction of the light, the light could never be extinguished merely by rotating a transparent film about a ray as axis. The Polaroid film A produces **plane-polarized** light, that is, the light emerging from A consists of vibrations occurring in one plane only. When B is perpendicular to A, therefore, no vibrations emerge from B.

Ordinary (*unpolarized*) *light* consists of vibrations occurring in *every* plane perpendicular to the ray.

Other methods of producing plane-polarized light. (1) *By reflection.* Polarized light is obtained when ordinary light is reflected from a plane sheet of glass or from a diffusing surface such as a wooden bench (Fig. 8.8*b*). When the angle of incidence is about 56° for glass of refractive index 1·5, all the reflected light is plane-polarized. This angle is called the "polarizing angle", and is given generally by $\tan i = n$ (Brewster's law).

(2) *By double refraction.* Crystals of calcite spar produce two beams of plane-polarized light from one incident beam of ordinary light. The vibrations in one beam are perpendicular to the vibrations in the other beam. This can be proved by rotating a Polaroid sheet, which cuts off each beam in positions at 90° to each other. One beam is said to consist of "ordinary" rays; the other of "extraordinary" rays.

Nicol prism. Used for producing and for testing plane-polarized light. It consists of a crystal of calcite spar of angles 72° and 108°, broken into two parts. The halves are cemented together with Canada balsam. The ordinary rays are totally reflected at the balsam; the extraordinary (plane-polarized) rays pass through.

To test whether light is plane-polarized, it is viewed through a Nicol prism or Polaroid which is rotated. If the light is completely extinguished at one stage, the light is plane-polarized.

Some uses of polarized light. (1) *Strains in structures* can be studied by viewing a model of the structure in polarized light. The patterns observed enable the engineer to calculate the strains, and to redesign the structure where necessary.

(2) *The concentration of sugar solutions* can be found by placing the solution between two "crossed" Nicols, that is, between two Nicol prisms through which no light emerges. With the solution present, however, light can now be seen, and one prism must be rotated through an angle θ to extinguish the light again. Experiments show that θ is proportional to the concentration, to a good approximation, and hence the concentration can be found. This method of testing sugar solutions is called "saccharimetry".

(3) *Polaroid glasses* reduce the glare of the sun reflected from a surface. They cut off the light due to vibrations perpendicular to the allowed direction of the Polaroid.

Sound

9. VELOCITY OF SOUND. FREQUENCY. INTENSITY

Sound waves. Material medium is required; sound does not pass through a vacuum. When a simple harmonic wave passes through air or any other medium, each layer vibrates with simple harmonic motion at the same frequency, but layers near one another are out-of-phase. At a given time, the displacements of the layers follow a simple harmonic pattern. See Fig. 9.1. A sound plane-progressive

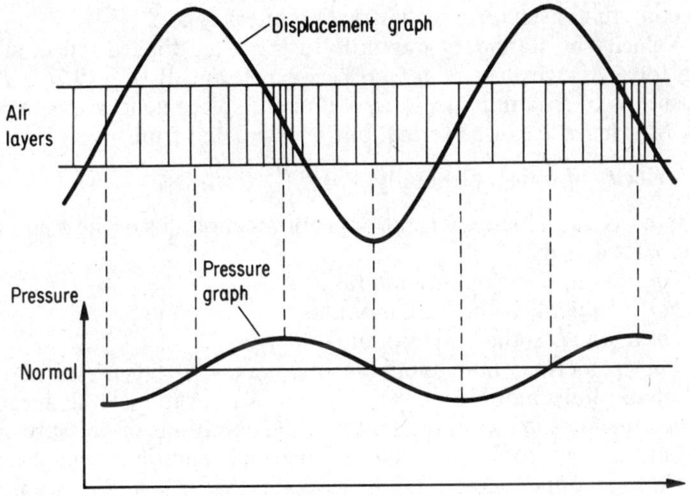

Particle displacement, $y = a \sin \frac{2\pi}{\lambda}(vt - x)$, where $v = f\lambda$

Wave velocity, $v = \sqrt{\frac{\gamma p}{\rho}}$

FIG. 9.1. Sound waves—displacement, pressure variations.

wave can be represented by the following expression for displacement y (see p. 54):

$$y = a \sin 2\pi\left(\frac{t}{T} - \frac{x}{\lambda}\right)$$

Particle displacement, y, varies with time t according to $y = a \sin \omega t$.

Particle velocity, v, varies with time according to $v\ (dy/dt) = \omega a \cos \omega t$, or with displacement y according to $v = \omega\sqrt{a^2 - y^2}$. Thus velocity is a maximum when $y = 0$.

Particle pressure varies in the same way as velocity. Thus pressure is a maximum (**compression** in wave) or a minimum (**rarefaction** in wave) where displacement is zero; pressure is normal where displacement is a maximum or a minimum (Fig. 9.1).

Velocity of sound in free air. Use of C.R.O. Join an audio-oscillator to a speaker L and to the Y_1-plates of a double-beam C.R.O. (Fig. 9.2*a*). Join a microphone M to the Y_2-plates of the C.R.O. Arrange the two sound waves to be in phase by moving L or M. Then move M a distance d until the two waves are again in phase. Taking the average of several values of d, this gives λ. Then $v = f\lambda$.

With a single-beam C.R.O., the wave obtained directly from L can be made to interfere with that received by M. The wavelength λ is then the distance between two successive positions of M for maximum constructive or destructive interference.

Velocity by stationary waves. By using a hardboard reflector B, stationary waves can be set up between L and B (Fig. 9.2*b*). The positions of the antinodes A (maximum displacement) are detected by M. Hence λ can be found, and v calculated from $v = f\lambda$.

Velocity of sound. Generally, $v = \sqrt{\dfrac{E}{\rho}}$,

where v is the velocity, E is the appropriate modulus of the medium, and ρ its density.

For a solid, E is Young's modulus.

For a liquid, E is the bulk modulus.

For a gas, E is the bulk modulus.

The isothermal bulk modulus of a gas $= p$, the *pressure*; the adiabatic bulk modulus $= \gamma p$, where γ is the ratio of the specific heat capacities. Newton thought that the variations of pressure and volume of a gas took place under isothermal conditions when transmitting a sound wave, and hence said that $v = \sqrt{p/\rho}$. Laplace reasoned that the changes were adiabatic and not isothermal, and corrected Newton's formula to the following:

$$v = \sqrt{\frac{\gamma p}{\rho}} \quad . \quad . \quad . \quad . \quad . \quad . \quad . \quad \text{(i)}$$

In the formula ρ is in kg m^{-3}, p is in N m^{-2}, and v is in m s^{-1}. Note that the pressure p due to 760 mm mercury, say, $= h\rho g$ $= 0{\cdot}76 \times 13600 \times 9{\cdot}81$ N m^{-2}.

Variation of velocity in a gas. For a given gas such as air:

(*i*) The velocity v is independent of the pressure,

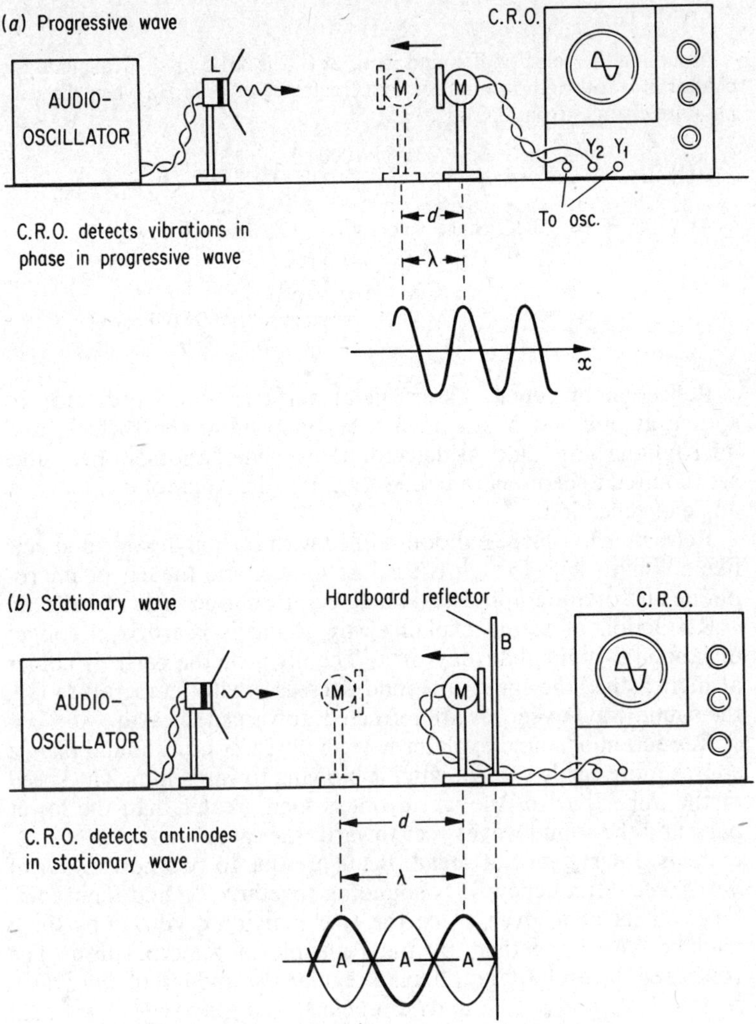

FIG. 9.2. Measurement of velocity of sound in free air.

(*ii*) The velocity $v \propto \sqrt{T}$, where T is the absolute temperature.

These statements are proved by substituting $\rho = m/V$ in (i), and then replacing pV by mRT. Since $v = \sqrt{\gamma RT}$, (*i*) and (*ii*) follow.

The presence of water-vapour in the air reduces the average density, since water-vapour is about $\frac{5}{8}$ths as dense as air. Thus generally the velocity of sound in moist air is greater than in dry air.

EXAMPLE

Assuming the velocity of sound in air at 0°C is 331·5 m s^{-1}, calculate the change in wavelength in mm due to a frequency of 1000 Hz when the temperature changes from 0°C to 27°C.

At 0°C, $\lambda_0 = \frac{v_0}{f} = \frac{331 \cdot 5 \times 1000}{1000} = 331 \cdot 5$ mm

At 27°C, $T = 300$ K. Since velocity $\propto \sqrt{T}$, new velocity $= \sqrt{300/273} \times 331 \cdot 5$ m s^{-1}.

$$\therefore \lambda_{27} = \frac{v_{27}}{f} = \sqrt{\frac{300}{273}} \times \frac{331 \cdot 5 \times 1000}{1000} = 347 \cdot 3 \text{ mm.}$$

Reflection of sound. Plane metal surface, audio oscillator as source at one end of an open tube pointing at the surface, and microphone with C.R.O. as detector at one end of another open tube, are required to demonstrate reflection. Result: Angle of incidence = angle of reflection.

Refraction of sound. Balloon is filled with carbon dioxide and acts like a convex lens. Watch is used as source, and the ear or microphone and oscillograph used as a detector on other side of balloon.

Refraction of sound explains why sound is heard over longer distances at night than day-time. The air near the earth is colder at night; since the speed of sound decreases with temperature fall, the sound waves veer, or are refracted, towards the earth.

Refraction of sound explains why an observer hears sound from a source more easily when a wind is blowing towards him. The speed of the upper part of the wavefront is then greater than the lower part, and the sound waves veer towards the earth.

Beats. Interference of sound. Beats are due to two notes A, B of *nearly equal* frequency f_1, f_2 sounding together. To find their combined effect at a given place the two individual vibrations there may be added together, by the Principle of Superposition. The combined or resultant graph has a regular rise and fall of *amplitude*, so that loud and soft sounds are heard. *The frequency of the loud sounds = the frequency of the beats* $= f_2 - f_1$ (Fig. 9.3).

To prove the formula, suppose t is the time between successive beats. In this time one fork makes exactly *one* cycle more than the other. Hence $f_2 t - f_1 t = 1$, or $f_2 - f_1 = 1/t$; but $1/t$ is the beat frequency. Thus $f_2 - f_1 =$ the beat frequency.

The combined effect of the two sources of sound produces intermittent *interference*. The resultant displacement, y, at a point $= a \sin \omega_1 t + a \sin (\omega_2 t + \theta)$, assuming the amplitudes a are equal.

$$\therefore y = 2a \sin \tfrac{1}{2}(\overline{\omega_1 + \omega_2 t + \theta}) . \cos \tfrac{1}{2}(\overline{\omega_1 - \omega_2 t - \theta})$$
$$= A \sin \tfrac{1}{2}(\overline{\omega_1 + \omega_2 t + \theta}),$$

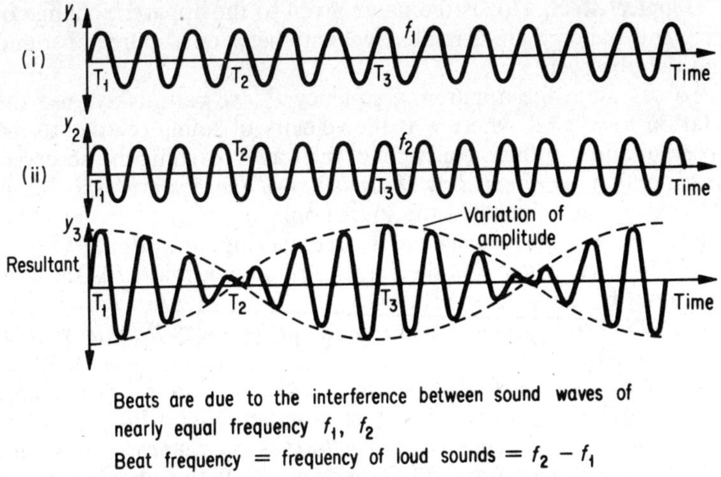

FIG. 9.3. Beats—interference of sound waves.

where $A = 2a\cos\frac{1}{2}(\overline{\omega_1 - \omega_2 t - \theta})$ = amplitude. By squaring A and expanding the result, we see that A^2 has a frequency given by $2\pi f = 2\pi f_1 - 2\pi f_2$, as $\omega = 2\pi f$ generally, or $f = f_1 - f_2$.

Besides its use in measurement of frequency, beats are used in tuning, for example, in the sonometer or resonance tube experiment.

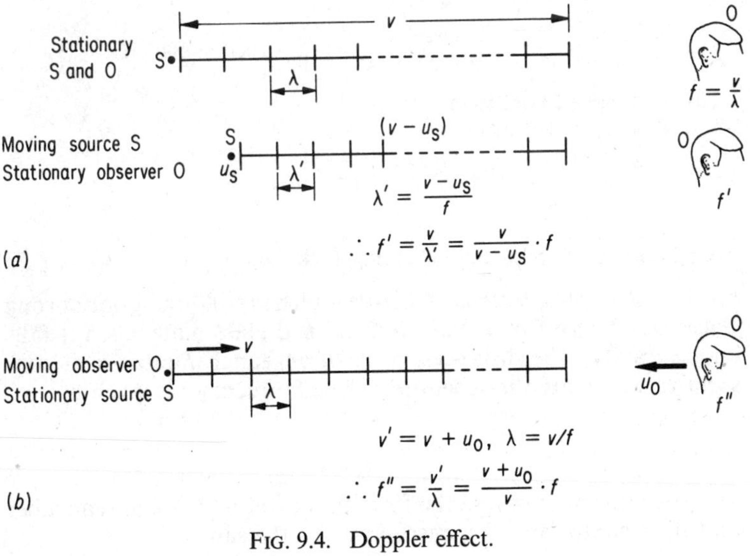

FIG. 9.4. Doppler effect.

Doppler effect. This is the name given to the apparent change of frequency when there is relative velocity between a source of sound and the observer.

To calculate the apparent frequency f', we can always use the relation $f' = v'/\lambda'$, where v' is the velocity of sound relative to the observer and λ' is the wavelength of the waves reaching the observer. In general, *v' is affected by the velocity of the observer only*; λ' is affected by the velocity of the source only.

(*i*) *Observer stationary and source moving towards it.* $v' = v$, $\lambda' = (v - u_s)/f$, where v is the velocity of sound in air, f is the true frequency of the source (Fig. 9.4*a*).

(*ii*) *Observer stationary and source moving away from it.* $v' = v$, $\lambda' = (v + u_s)/f$.

(*iii*) *Observer moving towards stationary source.* $v' = v + u_o$, $\lambda' = v/f$, where u_o is the velocity of observer (Fig. 9.4*b*).

(*iv*) *Observer and source moving towards each other.* $v' = v + u_o$, $\lambda' = (v - u_s)/f$. The apparent frequency in all the above cases is given by $f' = v'/\lambda'$.

EXAMPLE

A source of sound S, $f = 1000$ Hz, approaches a stationary observer O at 72 km h^{-1}. (*i*) Calculate the frequency heard by O. (*ii*) What is the frequency heard if S is stationary and O approaches S at 72 km h^{-1}? (Assume velocity of sound = 340 m s^{-1}.)

(*i*) v' (relative to observer) $= v = 340$ m s^{-1}.
λ' (reaching observer) $= (v - u_s)/f = (340 - 20)/1000$ m, since 72 km h^{-1} = 20 m s^{-1}.

$$\therefore f' = \frac{v'}{\lambda'} = \frac{340}{(340-20)/1000} = \frac{340}{320} \times 1000 = 1063 \text{ Hz.}$$

(*ii*) $v' = 340 + 20 = 360$ m s^{-1}.
$\lambda' = v/f = 340/1000$,

$$\therefore f' = \frac{v'}{\lambda'} = \frac{360 \times 1000}{340} = 1059 \text{ Hz.}$$

Measurement of frequency of tuning-fork

(1) *Falling-plate method.* A bristle attached to a tuning-fork prong traces out a series of waves on a smoked glass plate when it falls under gravity. The distances s_2, s_1 of an equal number n of consecutive waves are then measured. The frequency f is given by

$$= n\sqrt{\frac{g}{s_2 - s_1}}.$$

To prove this relation use the formula $s = ut + \frac{1}{2}gt^2$, and remember that the time to fall s_2, s_1 respectively is the same.

(2) *Stroboscope method.* Two light overlapping plates A, B, with slits S in each, are attached to the tuning-fork prongs (Fig. 9.5). The

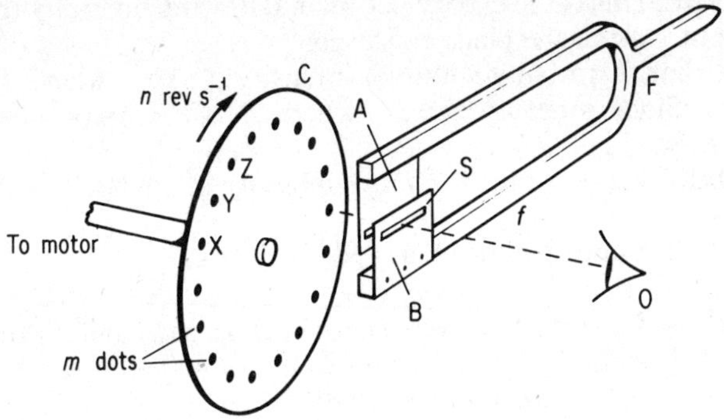

When dots X, Y, Z first appear stationary through S,

$$\text{time of successive glimpses} = \frac{1}{2} \times \frac{1}{f} = \frac{1}{m} \times \frac{1}{n}$$

$$\therefore f = \frac{mn}{2}$$

FIG. 9.5. Stroboscope method for f.

fork is electrically maintained. A white card C with a number of black dots m evenly spaced round the circumference is viewed through the slits, and the speed of the card is increased from zero. If the speed is n rev s^{-1} when the dots are first seen stationary, the frequency is given by $f = nm/2$.

As the speed is further increased, the dots appear to be moving forward. When the speed is $2n$ rev s^{-1} the dots are seen stationary again; there are half as many dots visible as before, as each dot has now moved two places forward between successive glimpses.

(3) *Beats.* Obtain beats with a fork of known frequency f_1 or with a siren. Suppose f_2 is the unknown frequency and f is the beat frequency. Then $f_2 = f_1+f$, or $f_2 = f_1-f$. To deduce which is correct, load the unknown fork with a little plasticine and count the beats again.

CHARACTERISTICS OF NOTES. SOUND INTENSITY

(1) The *pitch* depends on the frequency of vibration.

(2) The *loudness*, for a given observer, is proportional to the square of the amplitude of vibration and to the mass of air disturbed.

(3) The *timbre* or *quality* depends on the waveform, which is determined by the overtones accompanying the main (fundamental) note. The ear registers the overtones as a "background", and

distinguishes between the same note on different instruments as different overtones are obtained.

Intensity. This is the energy per second flowing normally through an area of 1 m^2 at the place concerned.

Note that, for a small source, intensity $\propto 1/r^2$, where r is the distance. Since intensity $\propto A^2$, where A is the amplitude, then $A \propto 1/r$. See Fig. 9.6*b*.

Decibel. This is a unit of *relative intensity* of sound. It is defined by

$$\text{number of decibels (dB)} = 10\log_{10}\left(\frac{P_2}{P_1}\right),$$

where P_1 is the original power (energy per second) and P_2 the final power of the sound. Thus a 40 milliwatt to a 400 milliwatt increase in power $= 10\log_{10}(400/40) = 10$ dB.

Intensity level. The lowest audible sound, the *threshold of hearing*, corresponds to an intensity of 10^{-12} watt metre^{-2} (P_0) (Fig. 9.6*a*). The *intensity level*, P, of a sound relative to P_0 as a "zero" is given by:

$$\text{intensity level} = 10\log_{10}\left(\frac{P}{P_0}\right)\text{ dB.}$$

Thus a loud sound of 10^{-1} W m^{-2} has an intensity level given by

$$10\log_{10}\left(\frac{10^{-1}}{10^{-12}}\right)\text{ dB} \quad\text{or}\quad 110\text{ dB.}$$

The *threshold of feeling* corresponds to the intensity of loud sound which causes a painful sensation.

Suppose a small loudspeaker produces an intensity level of 6 dB 40 m from it. At a distance 20 m from the loudspeaker, the intensity P_2 is four times the intensity P_1 40 m from it, since the intensity $\propto 1/r^2$ if the source is very small. Thus is P_0 is the reference level of intensity,

$$\textit{intensity level of } P_2 = 10\log_{10}\left(\frac{P_2}{P_0}\right) = 10\log_{10}\left(\frac{4P_1}{P_0}\right)$$

$$= 10\log_{10}4 + 10\log\left(\frac{P_1}{P_0}\right)$$

$$= 6 + 6 = 12\text{ dB.}$$

If the electrical power to the loudspeaker is reduced to one-third, the new intensity level 40 m from it is now given by

$$10\log_{10}\left(\frac{\frac{1}{3}P_1}{P_0}\right) = 10\log_{10}(\tfrac{1}{3}) + 10\log_{10}\frac{P_1}{P_0}$$

$$= -4{\cdot}8 + 6 = 1{\cdot}2\text{ dB.}$$

(Note the minus for the logarithm of a fraction.)

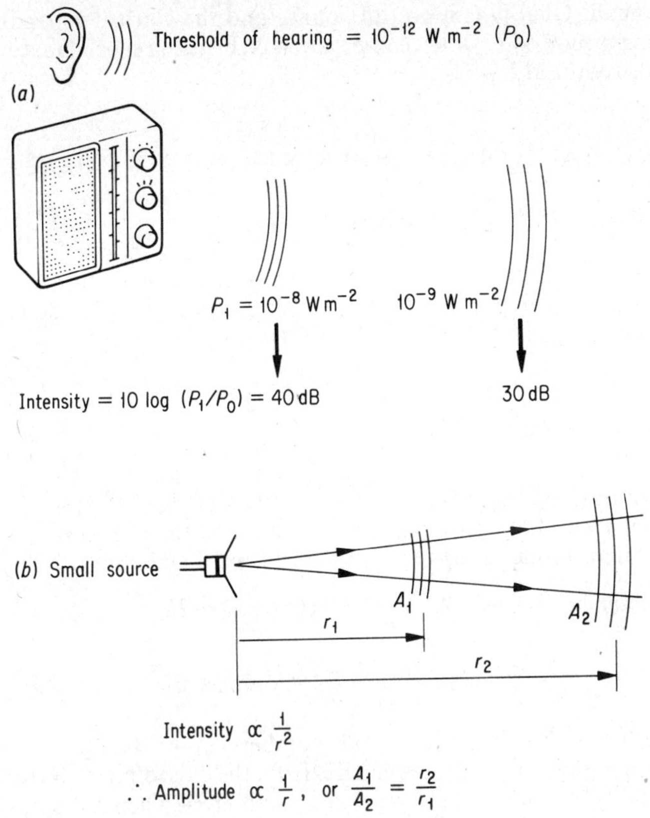

FIG. 9.6. Intensity of sound.

Phon. This is a unit of *loudness*. The source X whose loudness is required is placed near to a *standard source* Y of frequency 1000 Hz, and Y is then altered until its loudness is the same as X. Then if Y is *n* decibels above the threshold value (10^{-12} W m^{-2}), the loudness of X is said to be *n* phons. "Loudness" is subjective; it depends on the observer. "Intensity" is independent of the observer.

Reverberation. This is the name given to the perseverance of the sound in a room or hall after the source of sound ceases. The *time of reverberation T* was found by Sabine to be given approximately by

$$T = \frac{kV}{aA},$$

where V is the volume of the room, A its surface area, a the absorptive power of the surfaces, and k is a constant. Thus when a is high,

T is small. Quality (timbre) of sound, and the clarity, depends on the magnitude of T, which can be adjusted by using special materials for, or round, the walls.

10. VIBRATIONS IN PIPES. STRINGS. RODS

VIBRATIONS IN PIPES

Closed pipe. Stationary wave is formed, due to reflection at each end. A node always exists at the closed end, and an antinode at the open end (Fig. 10.1*a*).

Fundamental note. NA $= \lambda_0/4 = l$. Hence $\lambda_0 = 4l$.

$$\therefore f_0 = \frac{v}{\lambda_0} = \frac{v}{4l}.$$

Overtones. $3f_0, 5f_0, 7f_0, \ldots$ (odd numbers) (Fig. 10.1*a*).

Open pipe. Stationary wave is formed, due to reflection at each end. An antinode always exists at each (open) end (Fig. 10.1*b*).

Fundamental note. AA $= \frac{\lambda_0}{2} = l$. Hence $\lambda_0 = 2l$.

$$\therefore f_0 = \frac{v}{\lambda_0} = \frac{v}{2l}. \text{ (Compare the closed pipe.)}$$

Overtones. $2f_0, 3f_0, 4f_0, \ldots$ (all numbers) (Fig. 10.1*b*). Since the overtones have different frequencies from the closed pipe overtones, the quality or timbre of the same note is different when it is sounded on a closed and an open pipe.

Detection of nodes and antinodes in pipes. A membrane with fine sand particles is lowered into the pipe. Agitation of particles is considerable at the antinodes; the particles are at rest at the nodes.

The sensitive flame may be used to explore the pressure variation in the pipe. The flame flares up at a point of maximum pressure variation, which is a displacement node, and is quiescent where the pressure is normal, which is a displacement antinode.

End-correction of pipes. The antinode is a small distance c beyond the end of the pipe. Thus if l is the length of a *closed* pipe, $l+c$ = NA $= \lambda/4$ for the fundamental note, or $f_0 = v/\lambda = v/4(l+c)$.

For an *open* pipe, $l+2c = \lambda/2$, assuming the two end-corrections are equal.

Approximately, $c = 0{\cdot}6r$, where r is the radius of the pipe.

Resonance. Air in a pipe, or a taut string, or a wooden plank fixed at one end has a natural frequency, f_0, of vibration.

When a periodic force of frequency f acts on a mechanical system such as those just mentioned, it sets it into a *forced vibration*, of

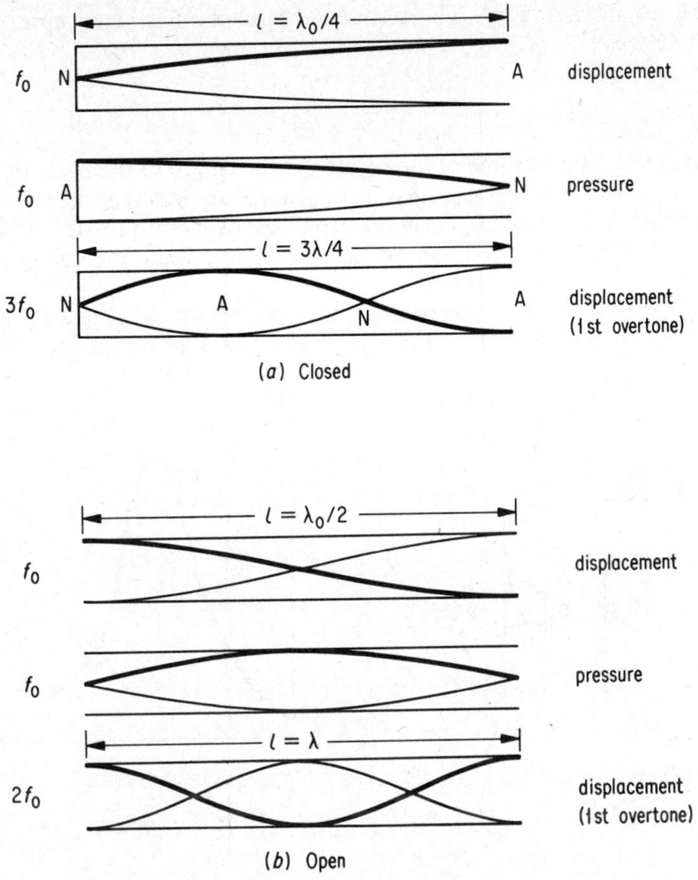

FIG. 10.1. Vibrations in (*a*) closed (*b*) open pipe.

relatively small amplitude when f is much different from the natural frequency f_0. When the frequency of the periodic force is f_0, the same as the natural frequency, it produces a very large amplitude of vibration, and the system is then said to be undergoing *resonant vibrations*. When the system vibrates by itself it undergoes *natural* or *free vibrations*.

The variation of amplitude of vibration as f varies is thus a maximum when $f = f_0$, and decreases when f becomes smaller or greater than f_0. See Fig. 10.2*c*.

With more friction present, i.e. more damping, the peak of the resonance curve becomes flatter.

Helmholtz resonator. When a note is obtained, the air in the neck of this resonator acts as a vibrating "piston" which stimulates the air in the vessel. The resonant frequency f is given approximately by:

$$f = \frac{1}{2\pi}\sqrt{\frac{\gamma p_0 a}{l \rho V}},$$

where V is the volume of the vessel, l is the length of the neck and a its area of cross-section, ρ is the density of air, γ is the ratio of the specific heats, and p_0 is the external atmospheric pressure. Thus $f \propto 1/v^{1/2}$, for a given narrow-neck resonator when the volume of air v below the neck is varied.

Resonance tube experiment. The water level is lowered until resonance occurs with a sounding tuning-fork of frequency f held

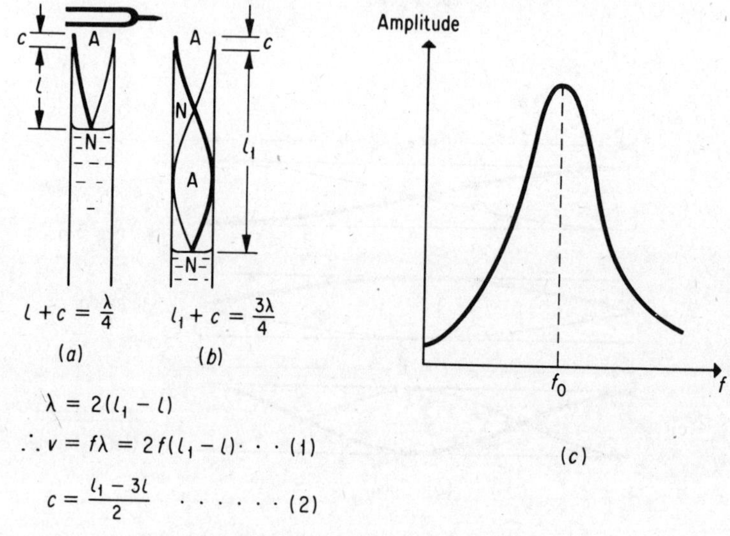

FIG. 10.2. Resonance tube experiment.

over the top. Then $l+c = \lambda/4$ (Fig. 10.2*a*). When the level is lowered further another position of resonance is obtained, corresponding to a length l_1 (Fig. 10.2*b*). Then $l_1+c = 3\lambda/4$.

$$\therefore\ l_1+c = 3(l+c), \quad \text{or} \quad c = (l_1-3l)/2 \quad . \quad . \quad . \quad \text{(i)}$$

Also, by subtraction to eliminate c,

$$l_1 - l = \frac{\lambda}{2}, \quad \text{or} \quad \lambda = 2(l_1-l)$$

$$\therefore\ v = f\lambda = 2f(l_1-l) \quad . \quad . \quad . \quad . \quad . \quad . \quad . \quad \text{(ii)}$$

Applications. (1) The end-correction can be determined. (2) The velocity of sound in a pipe can be determined. (3) The frequency f

of a tuning-fork can be found by measuring the first lengths, l, l_1 of resonance with the unknown fork and with a fork of known frequency f_1. Then, by knowing the end-correction c,

$$\frac{f}{f_1} = \frac{l_1+c}{l+c}.$$

(4) With various forks of frequency f and only *one* position of resonance, $l+c = \lambda/4 = v/4f$. A straight line graph is thus obtained when l is plotted against $1/f$. The slope of the line enables v to be found; the negative intercept of the line on the axis of $1/f$ gives c.

EXAMPLES

1. A vibrating tuning fork, of frequency 330 Hz, is held over a long glass tube full of water which is allowed to run out slowly. Calculate the lengths of air column giving the first two positions of resonance if the velocity of sound is 330·0 m s^{-1} and the diameter of the tube is 4·0 cm.

The end-correction, c, of the tube = 0·6 × radius = 1·2 cm.

Since $v = f\lambda$, the wavelength λ of the wave $= \frac{330 \times 100}{330} = 100$ cm.

At the first resonance position,

$$l+c = \frac{\lambda}{4}$$

$$\therefore l+1{\cdot}2 = \frac{100}{4} = 25$$

$$\therefore l = 23{\cdot}8 \text{ cm.}$$

At the next resonance position,

$$l_1+c = \frac{3\lambda}{4} = \frac{3 \times 100}{4}$$

$$\therefore l_1 = 75-1{\cdot}2 = 73{\cdot}8 \text{ cm.}$$

2. An open organ pipe sounding its fundamental note is in tune with a fork of frequency 439. How much must the pipe be shortened in order that two beats per second shall be heard when it is sounded with the fork? Assume the speed of sound is 342 metre per second.

When the pipe is shortened the frequency increases, to 439+2 or 441. The wavelength, λ, = $2l$ for the fundamental note, where l is the pipe length in cm.

$$\therefore f = \frac{v}{\lambda} = \frac{34200}{2l} = 439 \quad . \ . \ . \ . \ . \ . \ . \ . \ . \quad \text{(i)}$$

$$\therefore l = \frac{34200}{2 \times 439} = 39{\cdot}0 \text{ cm.}$$

Also, $$\frac{34200}{2l_1} = 441 \quad \ldots \ldots \ldots \quad \text{(ii)}$$

$$\therefore l_1 = \frac{34200}{2 \times 441} = 38{\cdot}8 \text{ cm.}$$

$\therefore l - l_1$ = shortening required = 0·2 cm.

VIBRATIONS IN STRINGS

Velocity of waves. When a taut horizontal flexible string is plucked, a *transverse wave* travels along it, that is, the vibrations of the particles of the string are vertical but the wave travels horizontally. The velocity, v, of the wave is given by

$$v = \sqrt{\frac{T}{m}},$$

where T is the tension and m is the mass per unit length. v is in metres per second when T is in newtons (N) and m in kg per metre. As shown later, the fundamental frequency of vibration f in a string of length l fixed at both ends is given by $f = \frac{1}{2l}\sqrt{\frac{T}{m}}$. The dependence on T and m (i.e. $f \propto \sqrt{T}$ and $f \propto 1/\sqrt{m}$) can be verified by the sonometer experiment, given below, and hence this is an indirect experimental verification of the velocity formula, $v = \sqrt{T/m}$.

When the string is stroked horizontally, a *longitudinal wave* travels along it, that is, the vibrations of the particles of the string are horizontal and the wave travels horizontally. The velocity, v, of this wave is given by

$$v = \sqrt{\frac{E}{\rho}},$$

where E is Young's modulus for the string and ρ is the density (see p. 54).

Frequency of waves. When a string is fixed at both ends, each end is a node, N, of the stationary wave (Fig. 10.3*a*). This applies to both transverse and longitudinal waves developed in the string. Thus NN = $\lambda/2 = l$, where l is the string length, or $\lambda = 2l$.

$$\therefore f = \frac{v}{\lambda} = \frac{1}{2l}\sqrt{\frac{T}{m}}, \text{ for transverse waves,}$$

and $$f = \frac{v}{\lambda} = \frac{1}{2l}\sqrt{\frac{E}{\rho}}, \text{ for longitudinal waves.}$$

Sonometer. When a string is plucked, then $f = \frac{1}{2l}\sqrt{\frac{T}{m}}$. Hence the frequency is inversely proportional to l, directly proportional to $\sqrt{T}$, and inversely proportional to $\sqrt{m}$.

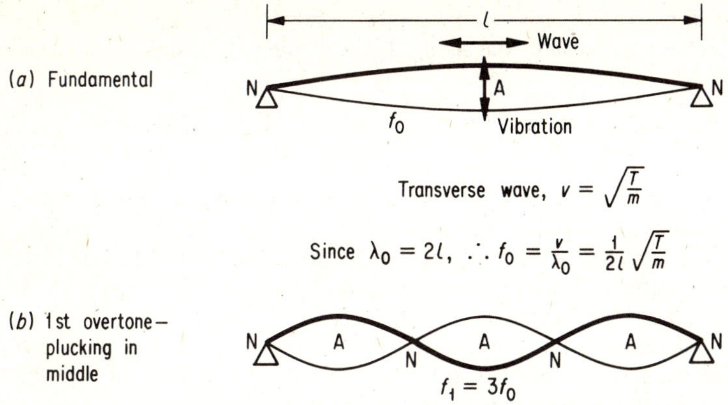

FIG. 10.3. Vibrations in strings.

To show $f \propto 1/l$, use five different tuning-forks and tune the length of wire on a sonometer, the tension being constant (Fig. 10.4*a*). Plot f against $1/l$, when a straight line passing through the origin is obtained.

To show $f \propto \sqrt{T}$, the wire XY of fixed length L and variable tension can be tuned to a second wire AB, having a fixed tension (Fig. 10.4*b*). The frequency of XY is then proportional to $1/l$, where l

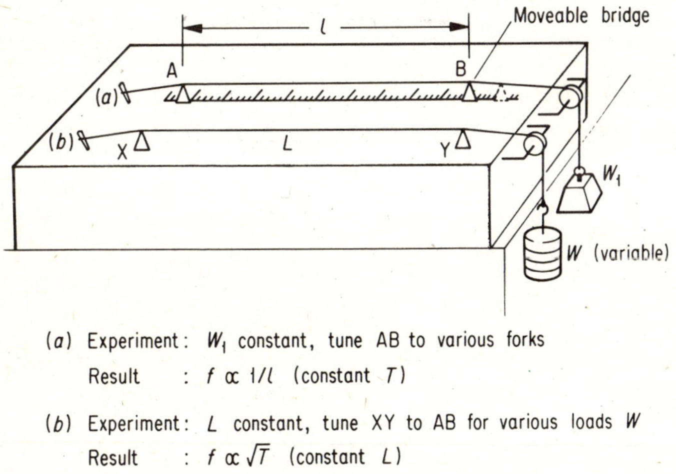

FIG. 10.4. Sonometer experiments.

is the length of wire AB. When $\sqrt{T}$ is plotted against $1/l(f)$, a straight line passing through the origin is obtained. Alternatively, the tension in XY can be altered until XY is in resonance with various tuning-forks and f plotted against $\sqrt{T}$.

Overtones of plucked string. If the string is plucked in the middle, this point is an antinode always. The frequencies obtainable are $f_0, 3f_0, 5f_0, \ldots$, where $f_0 = \frac{1}{2l}\sqrt{\frac{T}{m}}$ (Fig. 10.3*b*).

If the string is damped in the middle, this point is a node. The note obtained is $2f_0$. Other possible frequencies in this case are $4f_0, 6f_0, \ldots$

The overtones may be detected by means of a second wire B on a sonometer, having the same tension and material as the wire A. The length of B is arranged to be exactly one-third the length of A, and a small paper rider is placed at the mid-point of B. On plucking A the rider is observed to fall off, showing that a note of frequency $3f_0$ accompanied the note of frequency f_0 from A.

The frequency of the A.C. mains can be measured by connecting it to a sonometer wire AB, placing a horse-shoe magnet across the wire to obtain a perpendicular field, and then varying the tension, or

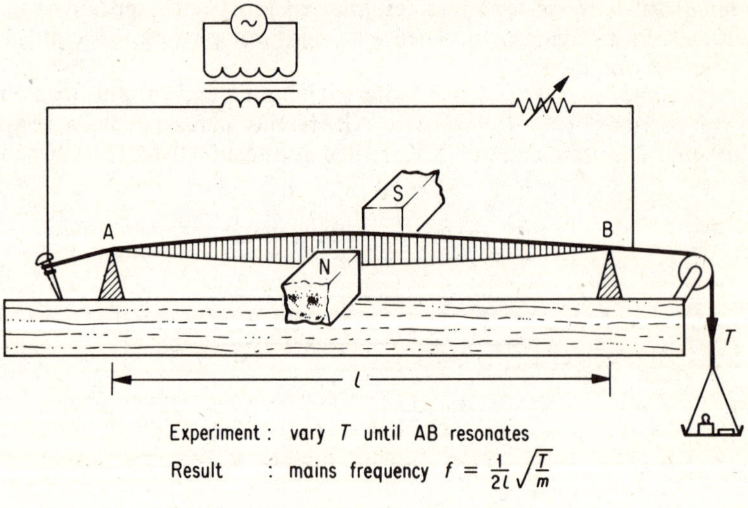

FIG. 10.5 Frequency of a.c. mains.

altering the length, of the wire until the wire is seen to be in resonance (Fig. 10.5). The frequency, f, of the mains is then given by

$$f = \frac{1}{2l}\sqrt{\frac{T}{m}}.$$

f is in Hz when l is in m, T in N, m in kg m^{-1}. *See Example* 2, p. 93.

Melde's experiment. When the prongs of a tuning-fork vibrate perpendicularly to a taut string and the tension T is varied, a number of loops n can be set up on the string. If f is the frequency of the fork,

$$f = \frac{n}{2l}\sqrt{\frac{T}{m}},$$

or $n^2 \propto 1/T$. This relation can be proved by varying T, observing n, and then plotting n^2 against $1/T$.

When the prongs of the tuning-fork vibrate in the same direction as the string,

$$\frac{f}{2} = \frac{n}{2l}\sqrt{\frac{T}{m}}.$$

EXAMPLES

1. A stretched wire is touched lightly with a feather at a point one-third of its length from one end, and is bowed near that end. It is found to be in unison with a fork of 512 Hz. Find the length of the wire, if the mass per metre length is 15×10^{-4} kg m^{-1}, and the tension is applied by a load of 10 kilogrammes force ($g = 9{\cdot}8$ m s^{-2}).

If the length of wire is l, then

$$\text{NN} = \frac{\lambda}{2} = \frac{l}{3}, \quad \text{or} \quad \lambda = \frac{2l}{3}$$

$$\therefore f = \frac{v}{\lambda} = \frac{\sqrt{T/m}}{2l/3} = \frac{3}{2l}\sqrt{\frac{T}{m}}$$

$$\therefore 512 = \frac{3}{2l}\sqrt{\frac{10 \times 9{\cdot}8}{15 \times 10^{-4}}}$$

$$\therefore l = \frac{3}{2 \times 512}\sqrt{\frac{10 \times 9{\cdot}8}{15 \times 10^{-4}}}$$

$$= 0{\cdot}75 \text{ m} = 75 \text{ cm}.$$

2. Two wires, one of phosphor bronze and the other of steel, are stretched between bridges 50 cm apart. An alternating electric current of frequency 50 Hz is passed along the bronze wire at the centre of which a permanent magnet is suitably placed, while an electromagnet energized by the same alternating current is placed at the centre of the steel wire which itself carries no current. Calculate the least tension required in each wire for it to vibrate in unison with the electrical supply, and determine what changes in tension would be necessary for the wires to remain vibrating if the supply frequency dropped by 0·5 Hz. [Diameter of phosphor bronze wire 1·0 mm, diameter of steel 0·8 mm. Density of phosphor bronze 8800 kg m^{-3}, density of steel 7800 kg m^{-3}.]

The wires vibrate in unison when their frequencies are 50 Hz respectively.

Now $f = \frac{1}{2l}\sqrt{\frac{T}{m}}$, where $l = 0{\cdot}5$ m, $m = \pi r^2 \times \text{density} = \pi \times (0{\cdot}4 \times 10^{-3})^2 \times 7800$ kg m^{-1} for steel.

(*a*) *Steel wire*:

$$50 = \frac{1}{2 \times 0{\cdot}5}\sqrt{\frac{T}{\pi \times 0{\cdot}4^2 \times 10^{-6} \times 7800}}$$

$$\therefore T = 50^2 \times 2^2 \times 0{\cdot}5^2 \times \pi \times 0{\cdot}4^2 \times 10^{-6} \times 7800$$

$$= 9{\cdot}8 \text{ N.}$$

(*b*) *Phosphor bronze wire*:

$$50 = \frac{1}{2 \times 0{\cdot}5}\sqrt{\frac{T}{\pi \times 0{\cdot}5^2 \times 10^{-6} \times 8800}}$$

$$\therefore T = 50^2 \times 2^2 \times 0{\cdot}5^2 \times \pi \times 0{\cdot}5^2 \times 10^{-6} \times 8800$$

$$= 17{\cdot}3 \text{ N.}$$

When the frequency drops to 49·5 Hz, the new tension T_1 required is given, since $f \propto \sqrt{T}$, by

$$\sqrt{\frac{T_1}{T}} = \frac{49{\cdot}5}{50}$$

For steel, $\therefore T_1 = \frac{49{\cdot}5^2}{50^2}T = \frac{49{\cdot}5^2}{50^2} \times 9{\cdot}8 = 9{\cdot}6$ N.

For phosphor-bronze, $T_2 = \frac{49{\cdot}5^2}{50^2} \times 17{\cdot}3 = 17{\cdot}0$ N.

VIBRATIONS IN RODS

Kundt's tube experiment. A rod AE is gripped at its mid-point. One end E has a light disc projecting into a closed glass tube T; the closed end R is adjustable (Fig. 10.6*a*). The inside of T is sprinkled lightly with lycopodium powder. On stroking A with a resined cloth and adjusting the length of T, the powder suddenly moves into a pattern and settles in heaps at regular intervals. *The powder demonstrates the existence of a stationary wave in the air.*

The ends A, E of the rod are antinodes, the middle is a node, N; thus the wavelength of the wave in the rod, $\lambda_r = 2l_r$, where l_r is the length of the rod. The wavelength, λ_a, of the wave in air $= 2l_a$, the distance between successive heaps. If v_r, v_a are the velocities of sound in the rod and air respectively, then, *since the frequency is the same,*

$$f = \frac{v_r}{\lambda_r} = \frac{v_a}{\lambda_a}.$$

$$\therefore v_r = v_a \cdot \frac{\lambda_r}{\lambda_a} = v_a \cdot \frac{l_r}{l_a}.$$

Measurement of velocities by Kundt's tube. The velocity in wood, glass, or a metal can thus be found by using a rod of the material,

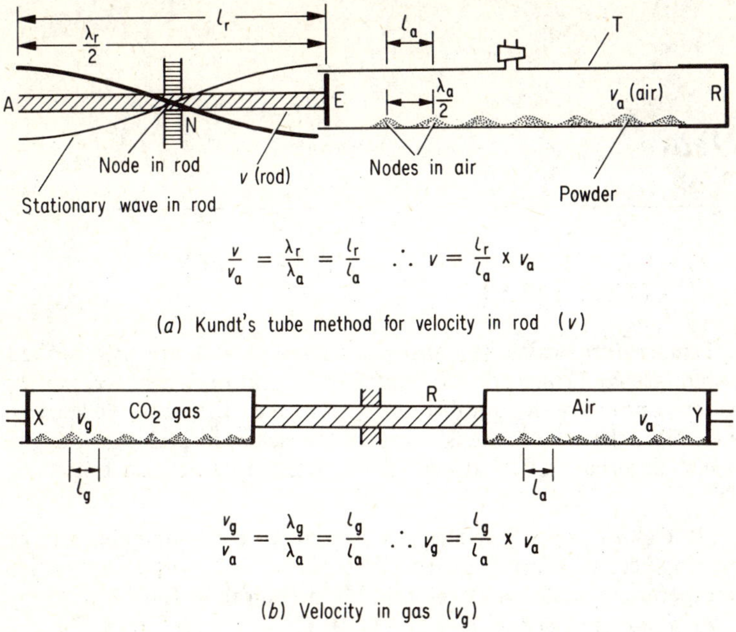

FIG. 10.6. Velocity by Kundt's tube.

measuring l_r and l_a, and using the velocity of sound in air, v_a, at the temperature concerned.

The velocity of sound in gases can be found by using a tube Y with air, and a tube X with the gas, at each end of the rod (Fig. 10.6*b*). Each tube is adjusted for resonance in turn. Then $v_g = v_a . l_g/l_a$, where l_g, l_a are the distances between successive heaps in X and Y respectively.

Young's modulus for a short thick bar can be found from the velocity of sound in it, using Kundt's tube method. Since

$$v_r = \sqrt{\frac{E}{\rho}},$$

$$E = v_r^2 \rho.$$

The ratio of the principal specific heats of a gas can be found from the velocity of sound in the gas. Since

$$v_g = \sqrt{\frac{\gamma p}{\rho}},$$

$$\gamma = v_g^2 \rho / p.$$

For a monatomic gas, $\gamma = 1{\cdot}67$; for a diatomic gas, $\gamma = 1{\cdot}4$. Molecular structure can thus be deduced from a Kundt's tube experiment.

Heat

11. THERMOMETRY. HEAT CAPACITY. LATENT HEAT

Temperature scales. (1) *Thermodynamic or absolute scale.* Fixed points are (i) Triple point of water (temperature where ice, water, water-vapour are in equilibrium) = 273·16 K, (ii) Absolute zero = 0 K. Note that this is the SI temperature scale (Fig. 11.1).

An "interval" of "1 degree K" is written "1 K", so that "per deg K" = K^{-1}.

(2) *Celsius scale.* Fixed points are: (i) Ice point (temperature of melting ice at 1 atmosphere pressure) = 0°C, (ii) steam point (temperature of steam at 760 mm Hg pressure) = 100°C.

Relations between K and °C. (1) 0°C = 273·15 K. (2) T K = (273·15 + t) K, where t is the temperature in °C. (3) 1 K (interval) = 1°C (interval), so that K^{-1} = $°C^{-1}$. See Fig. 11.1.

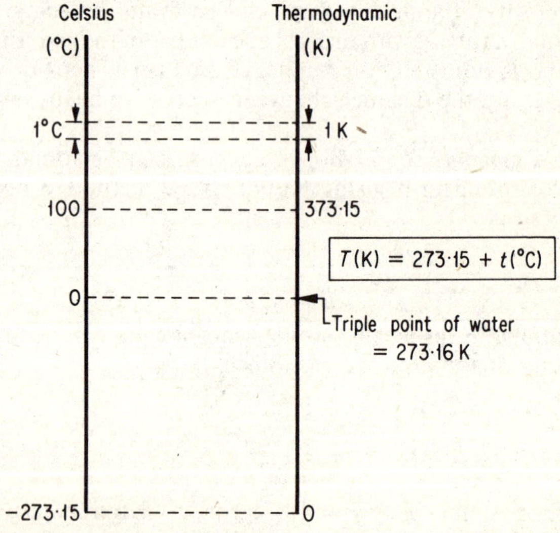

FIG. 11.1. Thermodynamic (Absolute) and Celsius scales.

Type	Fixed Points	Temperature Calculation	Range
Mercury-in glass thermometer	Volume of mercury, v_0, at melting ice temp. Volume of mercury, v_{100}, at steam temp. under standard pressure (760 mm mercury). —Celsius scale.	$t_m = \frac{v_t - v_0}{v_{100} - v_0} \times 100$ (Celsius)	−40°C to 300°C. Extended to 600°C by using nitrogen above mercury.
Platinum resistance thermometer	Resistance R_0, and resistance R_{100}, at above temperatures. —Celsius scale.	$t_{pt} = \frac{R_t - R_0}{R_{100} - R_0} \times 100$ (Celsius)	−200°C to 1000°C.
Gas thermometer (constant volume)	Pressure, p_{tr}, at triple point of water. (Thermodynamic)	$T = \frac{p_T}{p_{tr}} \times 273{\cdot}16$ (Kelvin)	−200°C to 1500°C (nitrogen). −250°C to 500°C (hydrogen).
Thermo-electric thermometer	E.m.f. E_0, and e.m.f. E_{100}, at Celsius fixed points.	$t_{th} = \frac{E_t - E_0}{E_{100} - E_0} \times 100$ (Celsius)	Copper-constantan (to 300°C). Chromel-alumel (to 1100°C). Platinum-platinum rhodium (to 1500°C). Lower range about −250°C.
Total radiation pyrometer	Thermocouple calibrated by reference to the gas thermometer scale.		Beyond 1500°C
Optical pyrometer	Electric lamp current related to temperature on gas thermometer scale.		700°C to beyond 1500°C.

Note. If the ice point is used, then T in K may be calculated for a gas-thermometer pressure p_T from the relation

$$T = \frac{p_T}{p_{ice}} \times 273{\cdot}15.$$

SOME DETAILS OF THERMOMETERS

Mercury-in-glass thermometer. Used domestically, but not for accurate work owing to (*i*) expansion of glass, (*ii*) exposed stem, (*iii*) non-uniformity of capillary, (*iv*) weight of mercury column, (*v*) "creep" of the zero.

Gas thermometer. *Temperature on this scale is used as a standard,* to which all others are referred. The simple constant volume air thermometer consists of a bulb connected to a mercury manometer, as in a Charles' law experiment (Fig. 11.2*a*). The accurate form of

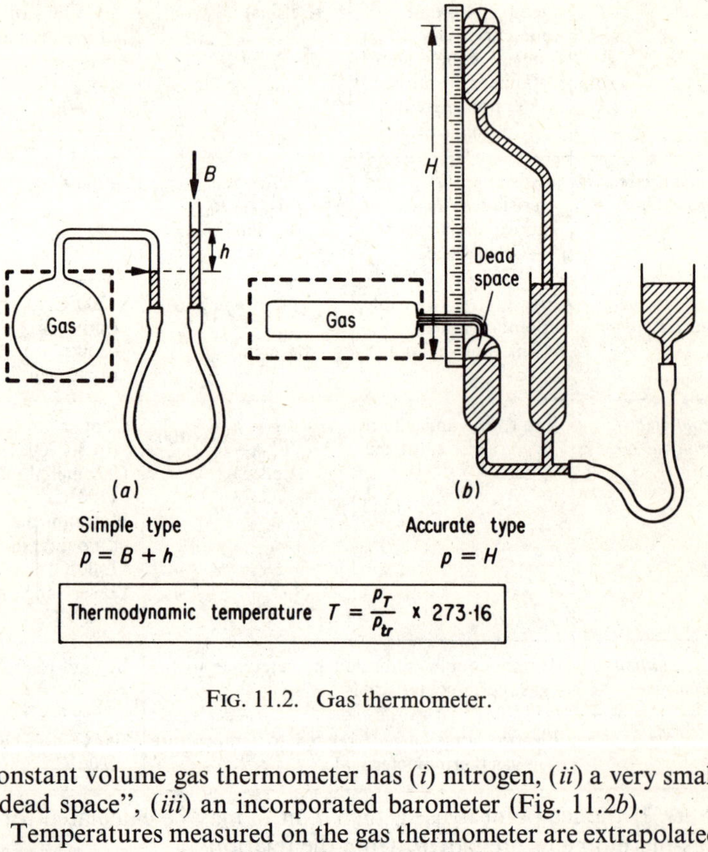

FIG. 11.2. Gas thermometer.

constant volume gas thermometer has (*i*) nitrogen, (*ii*) a very small "dead space", (*iii*) an incorporated barometer (Fig. 11.2*b*).

Temperatures measured on the gas thermometer are extrapolated to zero pressure, since all gases behave as ideal gases at zero pressure and the temperature readings are the same, whatever the nature of the gas.

Platinum resistance thermometer. Consists of platinum wire wound on mica frame inside a tube, with leads to two terminals at the top; "dummy" leads to two other terminals (Fig. 11.3). The leads resistance is eliminated by connecting the "dummy" set to the other side of a Wheatstone bridge circuit. Accuracy due to measurement of resistance to three decimal places. Widely used for steady liquid temperature. The temperature is "corrected" to gas thermometer scale by a standard formula.

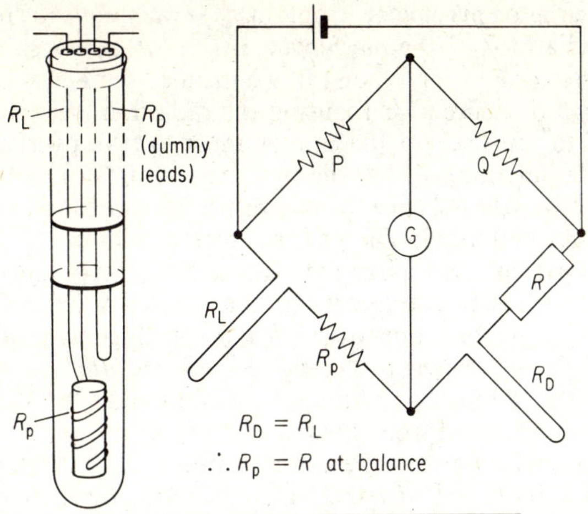

$R_D = R_L$

$\therefore R_p = R$ at balance

Celsius scale: $\dfrac{t}{100} = \dfrac{R_t - R_0}{R_{100} - R_0}$

FIG. 11.3. Platinum resistance thermometer and circuit ($P = Q$).

Thermo-electric thermometer. Copper-constantan (A–B) couple in glass tube. Other couples in porcelain for higher temperatures. Used for melting-points of metals, furnace temperatures, surface temperatures and varying temperatures. Time lag very small (Fig. 11.4).

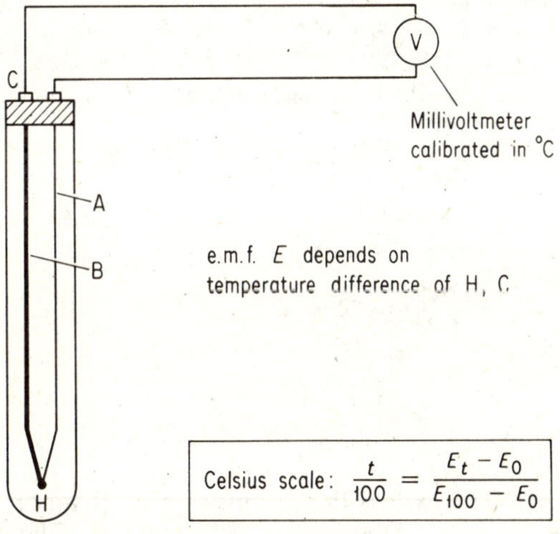

Celsius scale: $\dfrac{t}{100} = \dfrac{E_t - E_0}{E_{100} - E_0}$

FIG. 11.4. Thermoelectric thermometer.

Total radiation pyrometer. Contains (*i*) *concave mirror* for reflecting the radiation to (*ii*) a *thermopile*, (*iii*) a *galvanometer* reading in °C, (*iv*) an *eyepiece* sighted on (*v*) two inclined semi-circular *mirrors with an aperture*, used for focusing the radiation (see p. 145). The image of the furnace in the mirror must always overlap the hot thermopile junction, so that the pyrometer must not be placed too far away from the furnace. Temperature of sun found by using a rotating sectored disc in front of pyrometer (p. 144).

Optical pyrometer. Unlike the total radiation type, this one cannot be used for temperatures where the furnace or hot object is not glowing. Consists of a tube with (*i*) *lens* (providing image of furnace), (*ii*) *electric lamp filament*, (*iii*) *eyepiece with a red filter*. Accumulator, rheostat, and ammeter (graduated in °C) connected to filament, and reading observed when filament "disappears" (see p. 145).

International scale of temperature. Temperatures of freezing points and boiling points of oxygen, water, sulphur, silver and gold are determined on the constant volume gas thermometer scale, and are the fixed points on this scale. Other temperatures are found by using the platinum resistance thermometer and applying a formula.

EXAMPLE

The stem of a certain liquid-in-glass thermometer is marked correctly at 0°C and 100°C, and the length between the two marks is divided into 100 equal parts. What will be the reading on the scale so produced when the thermometer is immersed in a bath at 50°C if the volume of the liquid relative to glass is given by the expression $V = V_0(1+20\times10^{-4}t+3\times10^{-6}t^2)$ where t is the "correct" Celsius temperature?

The liquid volume at

$$50°\text{C}, V_{50}, = V_0(1+20\times10^{-4}\times50+3\times10^{-6}\times50^2);$$

$$\text{at } 100°\text{C}, V_{100}, = V_0(1+20\times10^{-4}\times100+3\times10^{-6}\times100^2).$$

The temperature t' on the liquid scale is given, by definition, by

$$t' = \frac{V_t-V_0}{V_{100}-V_0}\times100 = \frac{V_{50}-V_0}{V_{100}-V_0}\times100$$

$$= \frac{20\times10^{-4}\times50+3\times10^{-6}\times50^2}{20\times10^{-4}\times100+3\times10^{-6}\times100^2}\times100$$

$$= \frac{0{\cdot}1075}{0{\cdot}23}\times100 = 47°\text{C}.$$

HEAT CAPACITY. LATENT HEAT

Units of heat. The *joule* is the SI unit of heat. Symbol: J. 1 Kilojoule (kJ) = 10^3 J. 1 Megajoule (MJ) = 10^6 J. The joule was chosen as the unit of heat because heat is a form of energy.

The *calorie* (cal) is the heat required to raise the temperature of 1 g of water by 1°C. By definition, 1 cal_{IT} (international table) = 4·1868 J. Thus 4·2 J (approx.) = 1 cal.

The kilocalorie (kcal) is the heat required to raise the temperature of 1 kg of water by 1°C.

Heat (Thermal) capacity. This is the heat required to raise the temperature of an object by 1°C. Unit: J K^{-1}, or, if large, kJ K^{-1}. Symbol: *C*.

Specific heat capacity. This is the heat required to raise the temperature of unit mass of a substance by 1°C. Symbol: *c*. Unit: J kg^{-1} K^{-1}, or if large, kJ kg^{-1} K^{-1}. The specific heat capacity, c_W, of water is about 4·2 kJ kg^{-1} K^{-1} (4·2 J g^{-1} K^{-1}). Note that specific heat capacity is the "thermal capacity per unit mass", i.e. $c = C/m$, where m is the mass. Thus thermal capacity = mass × specific heat capacity.

Calculation. Find the heat needed to raise the temperature of 100 g of water in a copper calorimeter of mass 80 g, c = 0·4 kJ kg^{-1} K^{-1}, from 20°C to 70°C.

Water: Heat, Q_1, = 0·1 (kg) × 4·2 × 50 = 21 kJ
Copper: Heat, Q_2, = 0·08 × 0·4 × 50 = 1·6 kJ
∴ total heat = 22·6 kJ = 22600 J.

Methods for solids. (1) *Electrical heating*. In simple experiments, a solid metal block is heated electrically and the temperature rise θ observed after a time t (Fig. 11.5*a*). Then, assuming no heat losses,

$$IVt = mc\theta,$$

where IVt is the heat in joules, m the mass in kg and θ the temperature *rise* in °C.

(2) *Accurate electrical method* (Nernst and Lindemann). The metal itself is used as the calorimeter, and is heated by a platinum coil wound round the central block, from which the temperature is obtained. The metal is suspended in a vacuum. Then $IVt = mc\theta$, with the usual notation, where θ is temperature and t is time. This method is used to measure the variation of the specific heat of a metal (*i*) over a very small temperature range, (*ii*) at very low temperatures.

(3) *Mechanical energy method.* A solid metal calorimeter is rotated steadily. It rubs against a band with a constant tension F. The heat generated produces a temperature rise t. Then heat = $mct = F \times 2\pi rn$, where F is in newtons, r is the metal radius in metres, n is the number of revolutions. Thus c can be found (Fig. 11.5*b*).

(4) *Method of mixtures.* (Heat lost = heat gained; cooling correction needed.)

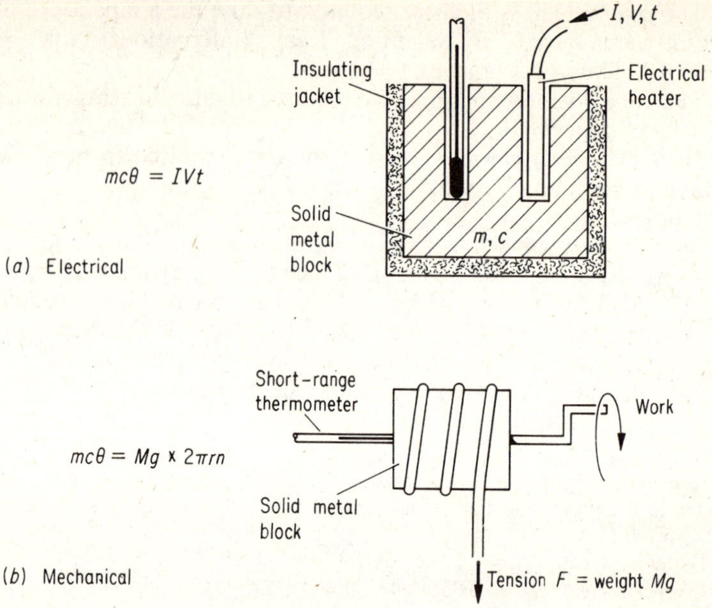

FIG. 11.5. Specific heat capacities.

(5) Bunsen's ice calorimeter or Joly's steam calorimeter (see below) used for *small amount of metal.*

(6) *Badly-conducting solid,* e.g. rubber. Method of mixtures and cooling correction, since solid takes a long time to give up its heat.

EXAMPLE

A block of metal of mass 500 g is heated electrically by a 100 W coil and reaches a constant temperature at 40°C. On switching off the heater, the rate of cooling is found to be $0{\cdot}2°C\ s^{-1}$. Calculate the specific heat capacity of the metal.

At 40°C, heat supplied per second = 100 W = $100\ J\ s^{-1}$.

At 40°C, heat lost per second = $mcd\theta/dt = 0{\cdot}5c \times 0{\cdot}2 = 0{\cdot}1c\ J\ s^{-1}$, where c is in $J\ kg^{-1}\ K^{-1}$.

Since the temperature is constant,

$$\therefore 0{\cdot}1c = 100, \text{ or } c = 1000\ J\ kg^{-1}\ K^{-1} = 1\ kJ\ kg^{-1}\ K^{-1}.$$

Bunsen ice calorimeter. Consists of a tube X surrounded by ice and air-free water in a vessel Y. Mercury fills the lower part of Y and also part of a horizontal capillary tube with a scale behind, so that the position of the mercury thread can be read.

When a mass m of pure water at t°C is poured into X, the thread contracts by a divisions, say. Thus a divisions are equivalent to a quantity of heat $mc_W t$. When a small amount of a solid, m_1 grams, at t_1°C is added, the thread moves back b divisions. Then $m_1 c t_1 = bmt/a$, from which c is found.

Theory. One gramme of ice occupies a volume of 1·09 cm^3 (approx.), since the density of ice is about 0·92 g cm^{-3}. Thus a contraction of 1·09 – 1 or 0·09 cm^3 occurs when 1 g of ice completely melts, i.e. when 334 J (approx.) of heat is given up to the ice. Thus a contraction of v cm^3 indicates that 334/0·09 joule is given up to the ice.

Joly's steam calorimeter. Consists of a metal sphere with collecting pans below it. The sphere is attached to the underside of a balance outside the chamber in which the sphere is situated. Steam is passed into the chamber, and some condensation, mass m say, takes place on the sphere. Shields prevent water from the roof falling on the sphere. Them ml, where l is the specific latent heat of steam, = the heat to raise the sphere and pans to steam temperature.

The experiment is repeated with a small quantity of solid in the pans below the sphere. More steam, m_1 say, now condenses on the sphere. Then

$$m_1 l - ml = m_2 c(100 - t),$$

where m_2 is the mass of solid, c its specific heat capacity, and t its original temperature, assuming 100°C to be the steam temperature. Thus c can be found.

Liquids. (1) *Constant-flow (electrical) method.* Liquid flows at steady rate through flow tube, due to K, and is heated electrically by coil R (Fig. 11.6). Reading of temperatures taken when they are steady. To eliminate heat lost per second by radiation, h, experiment is repeated with new rate of flow, and current adjusted to give same temperature rise $(t_2 - t_1)$ as before. Then, with the usual notation,

$$IV = mc(t_2 - t_1) + h,$$

$$I_1 V_1 = m_1 c(t_2 - t_1) + h.$$

Subtracting to eliminate h,

$$\therefore I_1 V_1 - IV = (m_1 - m)c(t_2 - t_1)$$

$$\therefore c = \frac{I_1 V_1 - IV}{(m_1 - m)(t_2 - t_1)}.$$

In the experiment, current and p.d. are measured by a potentiometer method, and t_2, t_1 by platinum resistance thermometers T_1, T_2. The accuracy of the method is due to (*i*) elimination of the unknown heat capacity of the flow-tube, (*ii*) current, p.d., and

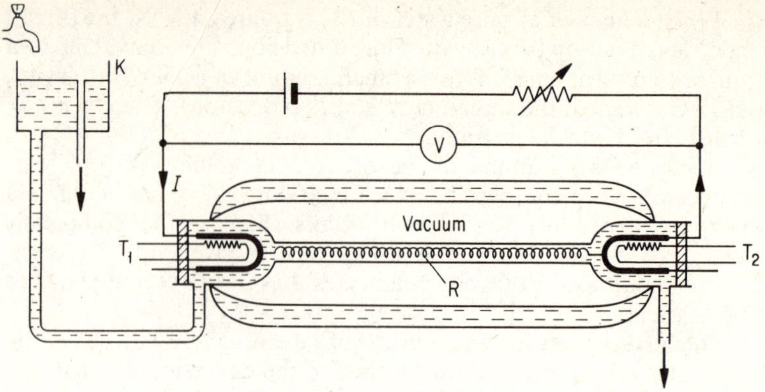

$$IV = mc(t_2 - t_1) + h$$

$$I_1V_1 = m_1c(t_2 - t_1) + h$$

$$\therefore c = \frac{I_1V_1 - IV}{(m_1 - m)(t_2 - t_1)}$$

FIG. 11.6. Specific heat capacity by constant flow tube.

temperature can be measured at leisure, (*iii*) electrical measurements can be made with high accuracy.

(2) *Variation of specific heat capacity*. Constant-flow tube method is used, with small rise of temperature, e.g. from 14°C to 16°C. The specific heat capacity is then the value at 15°C. The inlet temperature is now made 16°C and the outlet 18°C, and so on, increasing the inlet temperature each time. In this way the variation of the specific heat capacity of water with temperature is determined.

(3) *Method of mixtures,* assuming no heat of solution.

(4) *Method of cooling* for alcohol and other liquids which react with water. In this method (*i*) equal volumes of liquid and water are used in the same calorimeter, (*ii*) times for identical temperature drop are found from the cooling curves (Fig. 11.7). Since the average heat lost per second depends only on the area and the excess temperature over the surroundings, we have, with the usual notation,

$$\frac{C+mc}{t_1} \text{ (liquid)} = \frac{C+m_1c_W}{t_2} \text{ (water)}.$$

Cooling correction in method of mixtures. To eliminate the cooling correction, cool the liquid as much *below* room temperature as the final temperature will be *above* room temperature. Otherwise, the cooling correction, x°C, may roughly be found as follows: Let t seconds be the time for the liquid to heat up to its final observed temperature. Then x °C is the temperature *by which the liquid falls* in a time $t/2$ seconds.

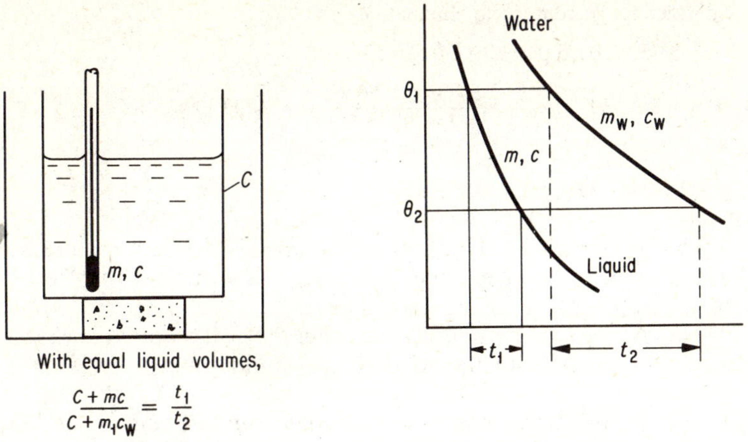

(*a*) Apparatus (*b*) Results

FIG. 11.7. Specific heat capacity by cooling method.

Newton's law of cooling. This states that *the rate of heat lost by an object is proportional to the excess temperature over the surroundings*. The law is true for all excess temperatures under conditions of forced convection; if there is no forced convection, the law holds only for small excess temperatures up to about 20°C. To verify the law plot a cooling curve, determine the rate of cooling by the *slope* of the curve at five temperatures, and plot the rate of cooling against the excess temperature concerned.

It should be carefully noted that this law is *not* used in the method of cooling for the specific heat capacity of a liquid, described p. 104.

EXAMPLE

Describe a continuous flow method of measuring the specific heat capacity of a liquid. Explain the advantages of the method.

Use the following data to calculate the specific heat capacity of the liquid flowing through a continuous flow calorimeter: *Experiment I.* Current 2·0 A, applied p.d. 3·0 V, rate of flow of liquid 30 g min^{-1}, rise in temperature of liquid 2·7°C. *Experiment II.* Current 2·5 A, applied p.d. 3·75 V, rate of flow of liquid 48 g min^{-1}, rise in temperature of liquid 2·7°C. (*L.*)

With the usual notation,

$$IV = mct + h,$$

where m is the mass of liquid per *second* flowing and h is the heat lost per second in J s^{-1}. Substituting the values given, and noting $m = 0{\cdot}5 \times 10^{-3}$ kg s^{-1} and $0{\cdot}8 \times 10^{-3}$ kg s^{-1} respectively,

$$2{\cdot}0 \times 3{\cdot}0 = 0{\cdot}5 \times 10^{-3} \times c \times 2{\cdot}7 + h \quad . \quad . \quad . \quad . \quad . \quad \text{(i)}$$

and

$$2{\cdot}5 \times 3{\cdot}75 = 0{\cdot}8 \times 10^{-3} \times c \times 2{\cdot}7 + h \quad . \quad . \quad . \quad . \quad . \quad \text{(ii)}$$

Subtracting (i) from (ii) to eliminate h,

$$2{\cdot}5\times 3{\cdot}75-2{\cdot}0\times 3{\cdot}0 = (0{\cdot}8-0{\cdot}5)\times 10^{-3}\times 2{\cdot}7c$$

$$\therefore c = \frac{2{\cdot}5\times 3{\cdot}75-2{\cdot}0\times 3{\cdot}0}{0{\cdot}3\times 10^{-3}\times 2{\cdot}7} = 4200 \text{ J kg}^{-1}\text{ K}^{-1}.$$

LATENT HEAT

The *specific latent heat, l,* of a substance is the heat required to change unit mass of it from one state to another without rise of temperature.

The *specific latent heat of fusion* is thus the heat required to change unit mass of it from the solid to the liquid state at the melting point.

The *specific latent heat of evaporation* (*vaporization*) is the heat required to change unit mass of it from the liquid to the vapour state at the boiling point.

Methods for specific latent heat of evaporation (vaporization). (1) *Water* (latent heat of steam). Steam trap and cooling correction required; final temperature not more than 30°C above surroundings; boiling-point "corrected" if barometric height is not 760 mm.

(2) *Electrical method.* Vapour passes down into a container at a steady rate when the liquid, heated electrically, begins to boil.

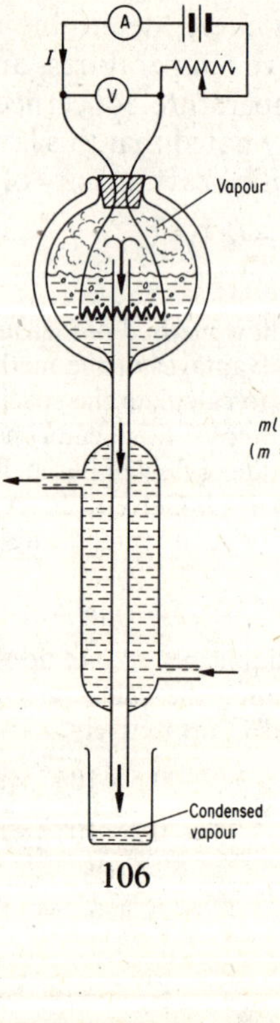

FIG. 11.8. Specific latent heat of vaporization.

Then $ml = IV$, where m is the mass of liquid collected per second (Fig. 11.8).

Variation of specific latent heat of evaporation. Variation of l with external pressure can be investigated by altering the pressure outside the liquid in the electrical method. When the pressure increases, l increases; the reverse is the case when the external pressure decreases.

Calculations. If a mass m of a vapour at its boiling-point θ°C condenses to liquid at a lower temperature t°C, then

$$\text{heat given up} = ml + mc(\theta - t),$$

where l is the specific latent heat of evaporation and c is the specific heat capacity of the liquid. Thus if 0·1 kg of steam at 100°C condenses to water at 30°C, then

$$\text{heat given up} = 0{\cdot}1\, l + 0{\cdot}1 \times c_W \times (100-30)$$

where l is the specific latent heat of steam.

Methods for latent heat of fusion. (1) *Ice*. Dry ice added to water. Final temperature must not be too low, otherwise it may be below the dew-point and water condenses from the atmosphere. To eliminate the "cooling correction", the water in the calorimeter can be warmed from room temperature, 15°C say, to 20°C (5°C above room temperature) and ice added until the temperature falls from 20°C to 10°C (5°C below the room temperature).

(2) *Metals* (Awbery and Griffiths). Large mass of metal was heated in a furnace until it melted, placed in a crucible, and drawn below water in a closed calorimeter. Final temperature measured. Crucible eliminated by another experiment.

12. EXPANSION OF GASES. KINETIC THEORY. SOLID AND LIQUID EXPANSION

EXPANSION OF GASES

Expansivity of a gas at constant pressure (volume coefficient)

$$\alpha_p = \frac{\text{increase in volume from 0°C}}{\text{volume at 0°C} \times \text{temp. rise}},$$

for a given mass of gas at constant pressure.

$$\therefore \alpha_p = \frac{V_t - V_0}{V_0 t},$$

or

$$V_t = V_0(1 + \alpha_p t). \quad \quad (1)$$

Pressure coefficient of a gas, $\alpha_V = \dfrac{\text{increase in pressure from 0°C}}{\text{pressure at 0°C} \times \text{temp. rise}}$, for a given mass of gas at constant volume.

$$\therefore \alpha_V = \frac{p_t - p_0}{p_0 \times t},$$

or $$p_t = p_0(1+\alpha_V t). \quad . \quad . \quad . \quad . \quad . \quad . \quad (2)$$

Charles' law states that a given mass of gas at constant pressure

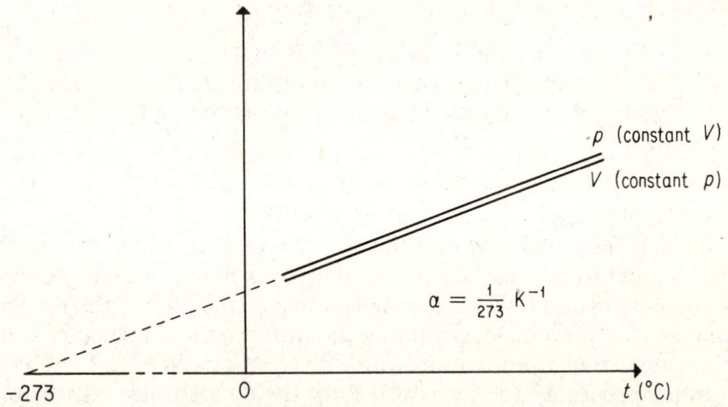

FIG. 12.1. Volume and pressure changes with temperature of ideal gas.

increases by 1/273 of its volume at 0°C for every °C rise in temperature (Fig. 12.1). Thus $\alpha_p = 1/273$. Hence, from (1)

$$\therefore V_t = V_0\left(1+\frac{t}{273}\right).$$

Expansivity = pressure coefficient. An *ideal gas* is one which obeys Boyle's and Charles' law. It is proved below that, for an ideal gas,

expansivity (volume coefficient) = pressure coefficient.

$$\therefore \alpha_V = \frac{1}{273},$$

or $$p_t = p_0\left(1+\frac{t}{273}\right).$$ See Fig. 12.1.

Proof: Start with the gas at 0°C and pressure p_0, volume V_0. Warm to t°C under (*i*) constant pressure, and (*ii*) constant volume. For (*i*), $V_t = V_0(1+\alpha_p t)$; for (*ii*) $p_t = p_0(1+\alpha_V t)$. From Boyle's law, $p_0 V_t = p_t V_0$; substituting for V_t, p_t, and eliminating $p_0 V_0$ on both sides, we find $\alpha_p = \alpha_V$.

Boyle's law. For a given mass of gas at a constant temperature, pV = constant.

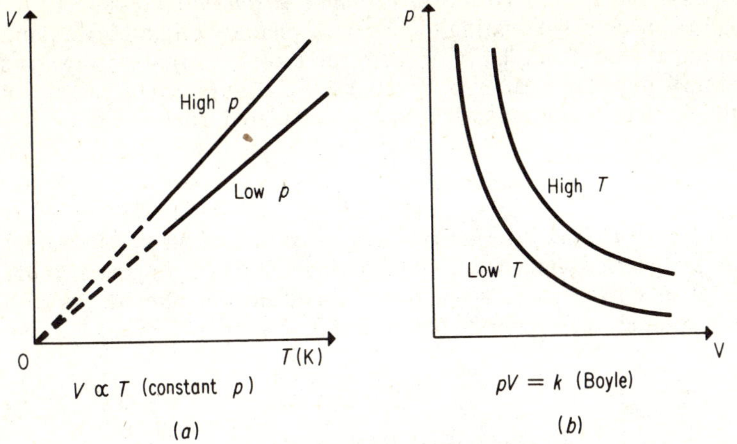

FIG. 12.2. Ideal gas laws.

Absolute temperature. Absolute temperature, T K = $273+t$°C. For a given mass of gas at constant pressure,

$$\text{volume} \propto T\ (\text{absolute}).$$

The volume of a gas *theoretically* becomes zero at 0 K or -273°C (approx.). See Fig. 12.2*a*.

For a given mass of gas at constant volume, pressure $\propto T$.

Ideal gas equation. Consider 1 kg of a gas at p_1, V_1, T_1, altered to p_2, V_2, T_2 in two stages: (*i*) Temperature constant at T_1, alter volume to V and pressure to p_2; then $p_2V = p_1V_1$. (*ii*) Pressure constant at p_2, alter temperature to T_2, and volume V to V_2; then $V/V_2 = T_1/T_2$. Thus

$$V = \frac{p_1V_1}{p_2} = \frac{T_1V_2}{T_2}$$

$$\therefore \frac{p_1V_1}{T_1} = \frac{p_2V_2}{T_2}$$

$$\therefore \frac{pV}{T} = R, \text{ a constant.}$$

$$\therefore pV = RT.$$

For any mass of a gas, $pV = mRT$, where R is the gas constant per kg, and m is the mass in kg.

For a mole of gas, $pV = \mathbf{R}T$, where **R** is a constant known as the *molar gas constant* since it is the same for one mole of any gas which occupies 22·4 litre at s.t.p. (0°C and 760 mm mercury pressure).

Experiments. (1) *To measure volume coefficient.* Pressure of air kept constant by equating the levels of mercury on both sides, and volume read from bulb of uniform cross-sectional area whose temperature can be altered. Volume at 0°C determined by producing the straight line graph back, and α_p calculated from

$$\alpha_p = \frac{V_t - V_0}{V_0 t}.$$

(2) *To measure pressure coefficient.* Volume of air kept constant by using a flask and connected manometer and bringing the mercury level to a fixed mark when the temperature was altered. As with the volume coefficient, the pressure at 0°C is obtained from a graph.

$$\alpha_V = \frac{p_t - p_0}{p_0 \times t}.$$

(3) *To measure temperature with simple air thermometer.* Using the pressure coefficient apparatus, measure the pressure (p_0) at 0°C, at 100°C (p_{100}) and at the unknown temperature t°C (p_t). Then

$$t = \frac{p_t - p_0}{p_{100} - p_0} \times 100.$$

If the gas law $p \propto T$ is assumed, then

$$\frac{p_t}{p_{100}} = \frac{273 + t}{373},$$

from which the unknown temperature t°C can be calculated when p_t, p_{100} are measured.

(4) *To verify* pV/T = *constant*, use the volume coefficient apparatus, and measure values of p, V, T for a given mass of air.

Magnitudes of Gas Constants. pV/T = a constant for a given mass of gas. The magnitude of the constant depends on the *mass* as well as on the nature of the gas.

Gas constant per kg. If we deal with 1 kg of the gas, the gas constant may be called R, and the gas equation is then written $pV = RT$. Suppose the density of hydrogen is 0·09 kg m^{-3} at s.t.p. (0°C and 760 mm mercury pressure). The volume of 1 kg is then 1/0·09 m^3. If the density of mercury is 13600 kg m^{-3}, standard pressure $p = h\rho g = 0{\cdot}76 \times 13600 \times 9{\cdot}8$ N m^{-2}.

$$\therefore R = \frac{pV}{T} = \frac{0{\cdot}76 \times 13600 \times 9{\cdot}8 \times 1}{273 \times 0{\cdot}09}$$

$$= 4{\cdot}15 \times 10^3 \text{ J kg}^{-1}\text{ K}^{-1} = 4{\cdot}15 \text{ kJ kg}^{-1}\text{ K}^{-1}$$

Molar gas constant. 1 mole of any gas has a volume of about 22·4 litres or $22{\cdot}4\times10^{-3}$ m³ at s.t.p.

$$\therefore \mathbf{R} = \frac{pV}{T} = \frac{0{\cdot}76\times13\,600\times9{\cdot}8\times22{\cdot}4\times10^{-3}}{273}$$

$$= 8{\cdot}3 \text{ J mol}^{-1}\text{ K}^{-1}$$

$$\text{Note that molecular weight of gas} = \frac{\text{molar gas constant, }\mathbf{R}}{\mathbf{R}\times10^{-3}}$$

General points. (1) When an open cylinder is inverted and submerged in water, Boyle's law can be applied to the air in the cylinder to calculate the level of water in it.

(2) If a faulty barometer contains air, Boyle's law can be applied to the air in problems on the height of mercury.

(3) When two bulbs are connected together, and one bulb is raised in temperature, the new pressure of air can be found by applying (*i*) *total* mass of air before heating = total mass after heating, (*ii*) mass of a gas = pV/RT. The new volume of air in the warm bulb at s.t.p. for example can be found from the new pressure value. (See Example below.)

(4) In the absence of melting ice, the simple constant volume gas thermometer can be used to measure temperature (*i*) by measuring the pressure p_1 at the temperature of boiling water, 100°C say, (*ii*) by measuring the pressure p at the unknown temperature t°C. Then, since pressure ∝ absolute temperature, t can be calculated from the relation

$$\frac{273+t}{373} = \frac{p}{p_1}.$$

EXAMPLE

State Boyle's and Charles' law and show how they may be combined to give the equation of state of an ideal gas.

Two glass bulbs of equal volume are joined by a narrow tube and are filled with a gas at s.t.p. When one bulb is kept in melting ice and the other is placed in a hot bath, the new pressure is 877·6 mmHg. Calculate the temperature of the bath. (*L.*)

First part. The equation of state of an ideal gas is $pV = RT$.

Second part. The mass of a gas, m, is given by $m = pV/RT$, where R is the gas constant for 1 kg. Thus if V is the volume of each bulb, then, since p = 760 mm and T = 273 K, the *total* mass of the gas, M, is given by

$$M = \frac{2\times760\times V}{R\times273}.$$

$$\text{The mass of gas in the bulb at } 0°\text{C} = \frac{877{\cdot}6 \times V}{R \times 273}.$$

$$\text{The mass of gas in the hot bulb at } T\text{ K say} = \frac{877{\cdot}6 \times V}{RT}.$$

Since the total mass of gas is constant,

$$\therefore \frac{877{\cdot}6V}{R \times 273} + \frac{877{\cdot}6V}{RT} = \frac{2 \times 76V}{R \times 273}.$$

Cancelling V and R,

$$\therefore \frac{877{\cdot}6}{T} = \frac{2 \times 760}{273} - \frac{877{\cdot}6}{273}$$

$$\therefore T = \frac{877{\cdot}6 \times 273}{642{\cdot}4} = 373 \text{ K}$$

$$\therefore \text{temperature} = 373 - 273 = 100°\text{C}.$$

KINETIC THEORY

Kinetic theory of gases. Basic assumptions: (1) The molecules are perfectly elastic, i.e. no energy lost on collision, (2) the time of collision is negligible compared with the time between collisions, (3) the volume of the molecules is negligible compared with the volume of the gas, (4) the attraction between the molecules is negligible. In these conditions, the gas is an *ideal gas*.

Calculation of pressure. Boyle's law. Consider molecules in the Ox direction moving towards a particular area A of the vessel normal to Ox (Fig. 12.3). The number of molecules per second

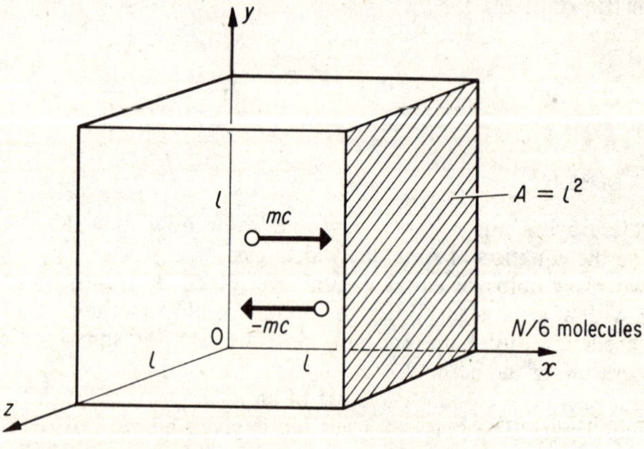

$2mc$ = momentum change per collision at wall

p = total mean pressure = $\frac{1}{3} N m\overline{c^2}/V$

FIG. 12.3. Kinetic theory of gases.

incident on $A = n_x Ac$, where n_x is the number of molecules per unit volume moving in the Ox direction and c is their average velocity.

$$\therefore \text{ force on } A = \text{momentum change per second}$$
$$= 2mc \times n_x Ac = 2n_x Amc^2,$$

where m is the mass of a molecule. Thus

$$\text{pressure, } p, = \frac{\text{force}}{A} = 2n_x mc^2 \quad . \quad . \quad . \quad . \quad . \quad . \quad . \quad . \quad \text{(i)}$$

Using $n_x = N/6V$, where N is the total number of molecules and V is the volume of the vessel, we obtain, from (i)

$$pV = \tfrac{1}{3}Nm\overline{c^2} = \tfrac{1}{3}M\overline{c^2}, \quad . \quad . \quad . \quad . \quad . \quad . \quad . \quad \text{(ii)}$$

where $\overline{c^2}$ is the *mean square velocity* and $M = Nm$ = mass of gas. The pressure p is thus proportional to $\overline{c^2}$, the mean square velocity.

From (ii), $$\sqrt{\overline{c^2}} = \text{r.m.s. velocity} = \sqrt{\frac{3p}{\rho}}, \quad . \quad . \quad . \quad \text{(iii)}$$

where ρ is the density of the gas. To calculate the r.m.s. velocity in m s^{-1}, p must be in N m^{-2} (if p = 760 mmHg, then $p = 1{\cdot}01325 \times 10^5$ N m^{-2}) and ρ in kg m^{-3}. From (ii),

$$pV = \tfrac{2}{3} \times \tfrac{1}{2}M\overline{c^2} \quad . \quad . \quad . \quad . \quad . \quad . \quad . \quad \text{(iv)}$$

$\therefore$ $pV = \frac{2}{3} \times$ *total kinetic energy of translation of molecules.*

Assuming the mean translational kinetic energy of the molecules depends only on the temperature, then pV = constant at the same temperature (Boyle's law). Since $pV = RT$, where R is the constant for the mass of gas, for example, 1 kg,

$$\text{total translational K.E. of gas} = \tfrac{3}{2}RT \quad . \quad . \quad . \quad \text{(v)}$$

Fig. 12.4 illustrates how translational kinetic energy, pressure, and the rate of diffusion, each depends on the velocity of the molecules.

EXAMPLE

Hydrogen has a density of 0·09 kg m^{-3} at 760 mmHg and 0°C. Find (*i*) the r.m.s. velocity of hydrogen at 27°C, (*ii*) the number of molecules of hydrogen which produce a pressure of 10^{-2} mmHg in a vessel of 1 litre volume at 0°C. (Density of mercury = 13600 kg m^{-3}, g = 9·8 m s^{-2}.)

(*i*) We have $p = 0{\cdot}760 \times 13600 \times 9{\cdot}8$ N m^{-2}.

$$\text{At } 0°\text{C}, \quad \therefore \quad \sqrt{\overline{c^2}} = \sqrt{\frac{3p}{\rho}} = \sqrt{\frac{3 \times 0{\cdot}76 \times 13600 \times 9{\cdot}8}{0{\cdot}09}}$$

Since r.m.s. velocity $\propto \sqrt{T}$, and 27°C = 300 K, 0°C = 273 K,

$$\therefore \text{ r.m.s. velocity at } 27°\text{C} = \sqrt{\frac{3 \times 0{\cdot}76 \times 13600 \times 9{\cdot}8}{0{\cdot}09}} \times \sqrt{\frac{300}{273}}$$

$$= 1930 \text{ m s}^{-1}.$$

(*ii*) At 760 mm and 0°C, the volume V occupied by the gas would be given in litres, from Boyle's law, by

$$760 \times V = 10^{-2} \times 1$$

$$\therefore V = \frac{10^{-2}}{760} \text{ litres} = \frac{10^{-5}}{760} \text{ m}^3$$

$$\therefore \text{ mass of gas} = \frac{0{\cdot}09 \times 10^{-5}}{760} \text{ kg.}$$

Now 2 kg of hydrogen, 1 kmol, contains $6{\cdot}03 \times 10^{26}$ molecules at 760 mmHg and 0°C. Hence if N is the number of molecules,

$$\frac{N}{6{\cdot}03 \times 10^{26}} = \frac{0{\cdot}09 \times 10^{-5}}{760 \times 2}$$

$$\therefore N = \frac{0{\cdot}09 \times 10^{-5} \times 6{\cdot}03 \times 10^{26}}{760 \times 2}$$

$$= 3{\cdot}6 \times 10^{17}.$$

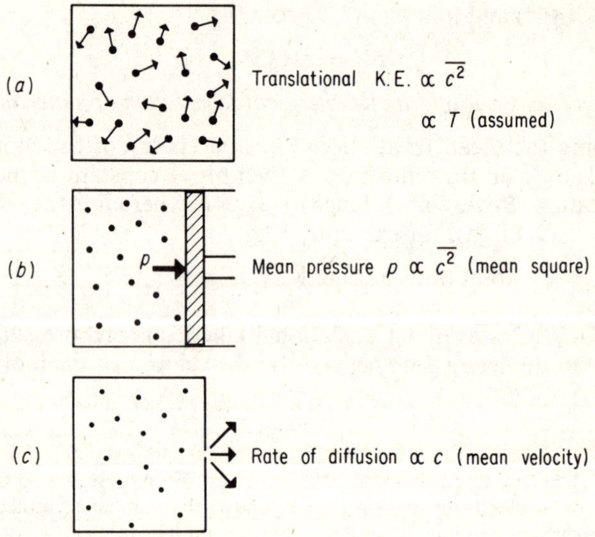

FIG. 12.4. Translational kinetic energy, pressure, diffusion

Distribution of velocities. The actual gas velocities are distributed among the molecules according to an exponential law due to Maxwell. The distribution is expressed by the relation $y = Ac^2e^{-\beta c^2}$, and A, β are constants which concern the number of molecules, their mass and nature, and their temperature. The number of molecules having velocities between c and $c+\delta c$ is $Ac^2 e^{-\beta c^2}.\delta c$.

More molecules have a velocity c_0 (at the peak) than any other value. c_0 = the most probable velocity, c_r = the root-mean-square velocity, c_m = the mean (average) velocity.

Graham's and Dalton's laws. From p. 113, r.m.s. value of gas velocity $= \sqrt{3p/\rho}$. Now a gas will flow through a fine hole at a rate depending on its mean velocity, which is proportional to its r.m.s. velocity (Fig. 12.4*c*).

$$\therefore \text{ rate of effusion of gas} \propto \frac{1}{\sqrt{\text{density}}}. \text{ (Graham's law.)}$$

Suppose we have a mixture of gases in a volume V, each exerting a pressure $p_1, p_2, \ldots$ respectively. Then, from (ii), p. 113,

$$p_1 V = \tfrac{1}{3}N_1 m_1 \overline{c_1^{\,2}}; \quad p_2 V = \tfrac{1}{3}N_2 m_2 \overline{c_2^{\,2}}; \ldots$$

Now average kinetic energy of the different gases is the same, since the temperature is the same, that is, $m_1\overline{c_1^{\,2}} = m_2\overline{c_2^{\,2}} = \ldots m\overline{c^2}$ say.

$\therefore$ $(p_1+p_2+\ldots)\ V = \frac{1}{3}(N_1+N_2+\ldots)m\overline{c^2} = \frac{1}{3}Nm\overline{c^2}$, where N is the total number of molecules.

$\therefore$ total pressure of gas $= p_1+p_2+\ldots$ = sum of partial pressures. (Dalton's law.)

Avogadro's law. Assuming no energy exchange between individual molecules of different gases brought together at the same temperature, then $\frac{1}{2}m_1\overline{c_1^{\,2}} = \frac{1}{2}m_2\overline{c_2^{\,2}}$. If the gases have equal pressures and volumes, $pV = \frac{1}{3}N_1 m_1\overline{c_1^{\,2}} = \frac{1}{3}N_2 m_2 \overline{c_2^{\,2}}$. Hence $N_1 = N_2$ (Avogadro's law).

Ratio of specific heat capacities. From (v), the energy E associated with three translational degrees of freedom $= 3RT/2$ for 1 kg of monatomic gas.

$$\therefore \text{ energy per degree of freedom} = \tfrac{1}{2}RT.$$

For a perfect *monatomic* gas, $c_V = dE/dT = \frac{3}{2}R$, $c_p = c_V + R = \frac{5}{2}R$

$$\therefore \gamma = \frac{c_p}{c_V} = \frac{5}{3} = 1{\cdot}67.$$

For a perfect diatomic gas, assuming 5 degrees of freedom (3 translational plus 2 rotational),

$$c_V = \tfrac{5}{2}R, \quad c_p = c_V + R = \tfrac{7}{2}R.$$

$$\therefore \gamma = \tfrac{7}{5} = 1{\cdot}40.$$

Generally, $\gamma = 1 + \dfrac{2}{n}$, where n is the total number of degrees of freedom.

Evidence for Kinetic Theory. (1) γ, measured by Clément and Desormes' experiment (p. 126), is about 1·67 for a monatomic gas, as predicted, (2) coefficient of viscosity, η, of a gas is independent of pressure for moderate values, as predicted, (3) Brownian motion experiments by Perrin show that the number of molecules in a kilo-mol (kmol) of liquid is of the order 6×10^{26}, the Avogadro constant, the same as for a gas.

Temperature measurement of hot gas. Owing to the Doppler effect (p. 82), the breadth of a spectral line of wavelength λ is $2\lambda v/c$, where v is the velocity of the gas molecules and c is the velocity of light. Since $v = \sqrt{3RT}$ (approx.), the temperature T can be found from measurements of the breadth of the spectral line.

Real gases. Deviations from the gas laws. (1) For a real gas, the pressure inside the gas $= p+a/V^2$, where a/V^2 is a measure of the attractive force on the molecules striking the wall of those behind and p is the observed pressure. Since the "effective" volume is $(V-b)$, where b is a factor depending on the dimensions of the molecules, we have

$$\left(p+\frac{a}{V^2}\right)(V-b) = RT.$$

This is known as *van der Waals' equation*. Andrews' results, for carbon dioxide in Fig. 13.7, p. 132, show that real gases have a cubic form of $p-V$ relation, similar to that predicted by van der Waals' equation. Fig. 12.5*a* shows p v. V curves for real and ideal gases, the latter obeying Boyle's law.

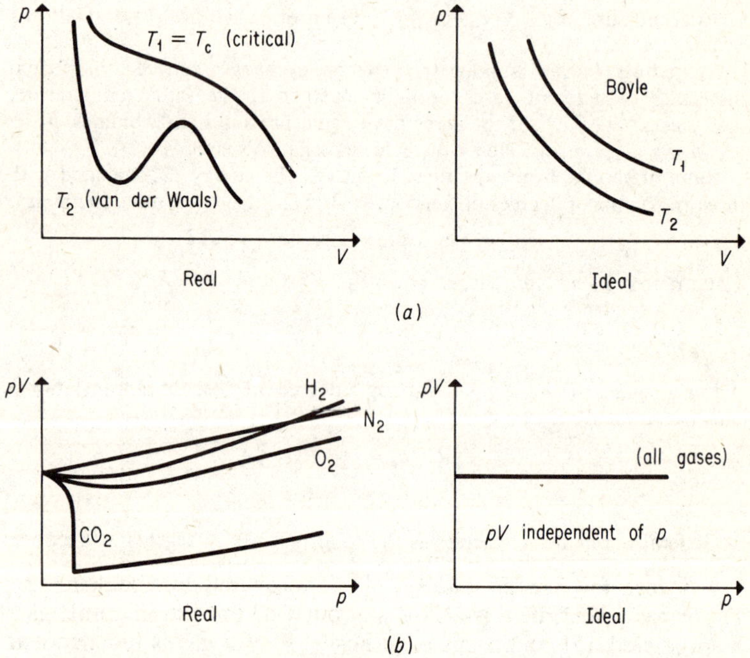

FIG. 12.5. Real and ideal gases: (*a*) p against V (*b*) pV against p.

(2) The graph of pV, v, p should be a straight line parallel to the pressure-axis, if the gas obeyed Boyle's law and was an ideal gas. Experiments by Amagat at high pressures show that for air, carbon dioxide and nitrogen, pV first decreases as p increases,

reaches a minimum, and then increases. For hydrogen, pV increases as p increases. Generally, $pV = A+Bp+Cp^2+\ldots$, where $A = RT$. Thus a gas obeys Boyle's law only at very low pressures. Fig. 12.5*b* shows pV, v, p curves for real and ideal gases; for the latter, pV is independent of p.

THERMAL EXPANSION OF SOLIDS AND LIQUIDS

Solids

Molecular theory of expansion. On heating a solid, the vibrational energy of the molecules increases and so does the intermolecular potential energy V (see p. 147). The mean separation r between molecules is now greater than the separation r_0 before heating. Fig. 15.1, p. 146, illustrates the variation of V with r. Since the curve is not symmetrical, the mean separation at a higher energy level XY is G, which is to the right of C′. On heating, then, $r > r_0$; and so expansion occurs.

Linear expansivity, α, is the increase in length per unit length at 0°C per degree rise in temperature. Unit: K^{-1}.

Mean linear expansivity between two temperatures t_1°C, t_2°C

$$= \frac{\text{increase in length}}{\text{original length} \times (t_2 - t_1)}.$$

Laboratory experiment for linear expansivity. (*i*) Micrometer screw gauge, or optical lever, for measuring the small increase in length. (*ii*) Heating jacket with steam round a rod of the material. (*iii*) Measurements, at steady conditions: Change in length, original length of bar, initial temperature, final temperature.

Accurate (comparator) method for linear expansivity. Two scratches X, Y made on the bar B concerned, about 1 metre apart (Fig. 12.6). The bar is placed in a bath containing melting ice, and the scratches are each observed through a telescope M with an eyepiece containing a vernier scale. Length between scratches obtained by using a standard metre, which is wheeled away. The bath is then warmed to a definite temperature. The small increase in length is measured from the shift of each scratch.

The telescopes are each attached to massive concrete pillars, so that their position is independent of temperature. In laboratory experiments, however, slight changes in the position of the screw-gauge may occur when the temperature alters in its neighbourhood.

Expansion of small solid or crystal. Interference bands counted when solid expands and air film thickness decreases. Allowance made for upward expansion of platform on which solid rests.

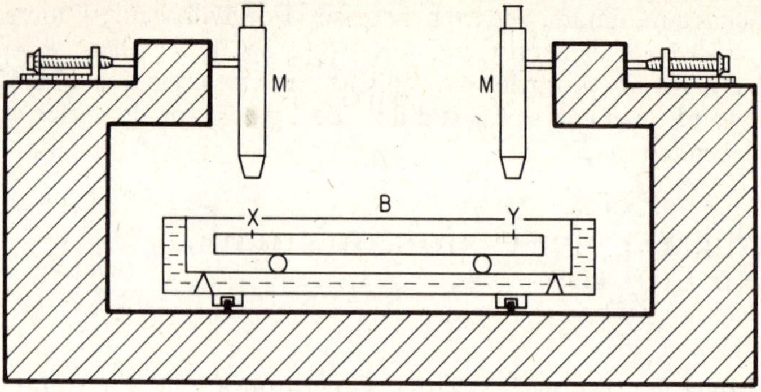

FIG. 12.6. Linear expansivity by comparator method.

Formulae. Increase in length $= \alpha lt$, where l = original length, t = temperature rise, α = mean expansivity.
New length, l_t, $= l(1+\alpha t)$.
Superficial (area) expansivity $= 2\alpha$;
cubic (volume) expansivity $= 3\alpha$.

General points. (1) Force in bar due to expansion or contraction $= EA\alpha t$, where E = Young's modulus, A = cross sectional area, t = temperature rise.

Units: E in N m^{-2}, A in m^2, t in °C (or K), α in K^{-1}.

EXAMPLE

An iron rod of cross-sectional area 72 mm^2 is heated so that its temperature is increased by 100°C. What force will be required to hold the rod at its expanded length when the rod is cooled to its original temperature? (Assume linear expansivity of iron $= 1{\cdot}1 \times 10^{-5}$ K^{-1}, Young's modulus $= 2{\cdot}0 \times 10^{11}$ N m^{-2}.)

$$\text{Force, } F = EA\alpha t$$
$$= 2 \times 10^{11} \times 72 \times 10^{-6} \times 1{\cdot}1 \times 10^{-5} \times 100$$
$$= 15840 \text{ N.}$$

(2) The true length at 16°C when 774 mm is read on a scale correct at 0°C $= 774(1+16\alpha)$ mm.

(3) The new period T_2 of a pendulum bob suspended by a wire of linear expansivity $\alpha = T_1\sqrt{1+\alpha t}$, where T_1 is the old period and t is the temperature rise ($T = 2\pi\sqrt{l/g}$).

Liquids

Absolute cubic expansivity, γ = true increase in volume per unit volume at 0°C per unit temperature rise.

$$\text{New volume, } V_t = V_0(1+\gamma t).$$

Apparent cubic expansivity, γ_{app} = the apparent increase in volume per unit volume at 0°C per unit temperature rise (i.e. the expansion relative to the container expansion). Approximately,

$$\gamma = \gamma_{app}+g,$$

where c is the *cubic expansivity* of the container's material.

Density relation

$$\frac{\rho_0}{\rho_t} = 1+\gamma t,$$

where ρ_0, ρ_t are the densities at 0°C, t°C respectively. Note that ρ_0 is greater than ρ_t usually, and the density ρ_1 at t_1°C is greater than the density ρ_2 at t_2°C if t_1 is less than t_2. The density relation is proved by using mass $= V_0\rho_0 = V_t\rho_t$.

Dulong and Petit's (balancing column) method for absolute expansivity. Utilizes the formula: "pressure$=h\rho g$". The method provides the absolute cubic expansivity value directly because it is independent of the expansion of the vessel (pressure is independent of cross-sectional area and depends only on height).

Principle. In the simplest form of apparatus, one side of a U-tube is heated to steam temperature, the other side is kept at 0°C. Starting with: "Pressure is the same at the same horizontal level of a liquid" and equating the pressures on both sides at the bottom, we obtain $B+h_0\rho_0 g = B+h_1\rho_1 g$, or $h_0\rho_0 = h_1\rho_1$. Using $\rho_0/\rho_1 = 1+\gamma t$, we find

$$\gamma = \frac{h_1-h_0}{h_0 t}.$$

The distance (h_1-h_0) is measured by a travelling microscope; h_0 may be measured by metre rule. Defects of the simple Dulong and Petit experiment are: (*i*) small difference in levels, (*ii*) wide separation of levels, (*iii*) conduction and convection from "hot" to "cold" side, (*iv*) "corrections" needed for mercury thermometers, (*v*) surface tension affects the meniscus level.

Laboratory apparatus. In one laboratory form, the two open ends are brought together to make measurement of (h_1-h_0) more easy and accurate. In Fig. 12.7, the "cold" side and h_2 are all at room temperature, and if t is the temperature difference between the two sides,

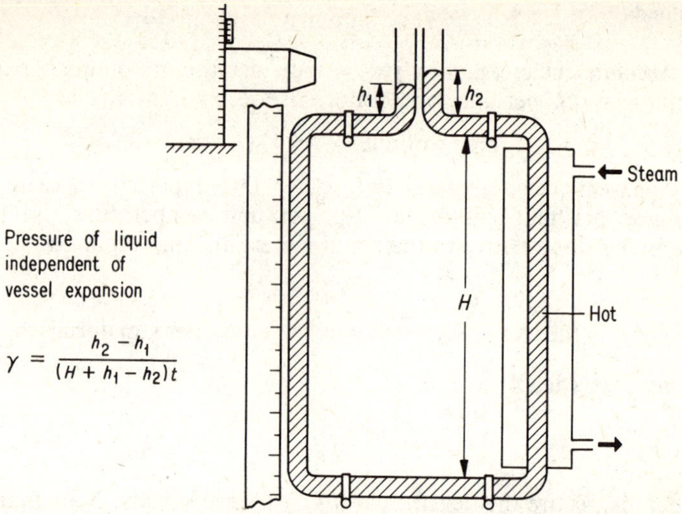

FIG. 12.7. Absolute expansivity by pressure method (Dulong and Petit).

$$\gamma = \frac{h_2 - h_1}{(H + h_1 - h_2) \times t}.$$

Accurate apparatus. Callendar and Moss measured the expansivity of mercury accurately (*i*) by using six pairs of "hot" and "cold" columns, which increased the difference in levels, (*ii*) by using melting ice on one side and a warm liquid on the other, with platinum resistance thermometers for temperature measurement. In general, γ may be calculated from

$$\frac{h_2 - h_1}{1 + \gamma t_1} + \frac{H_1}{1 + \gamma t} = \frac{H_0}{1 + \gamma t_2} + \frac{h_3}{1 + \gamma t_3},$$

where t_1, t_2, t_3 are the "cold" temperatures, t is the "hot" temperature, H_0, h_1, h_2, h_3 are the "cold" column lengths, and H_1 is the "hot" column length.

Method for apparent expansivity

(1) *Density bottle method, or weight thermometer.* This gives the apparent expansivity:

$$\gamma_{app} = \frac{\text{mass expelled}}{\text{mass LEFT} \times \text{temp. rise}}.$$

Absolute expansivity, $\gamma = \gamma_{app} + g$ (approx.),
where g is the cubic expansivity of the vessel's material (3 times the linear expansivity).

To prove the formula for γ_{app}, begin with (*i*) $m_1 = V_1 \rho_1$ (t_1°C),

$m_2 = V_2\rho_2\,(t_2°C)$; (*ii*) then divide the two equations, i.e. $m_1/m_2 = (V_1/V_2)\times(\rho_1/\rho_2)$; (*iii*) substitute $1+\gamma t$ for ρ_1/ρ_2, and $1/(1+gt)$ for V_1/V_2; (*iv*) cross-multiply and simplify for γ.

(2) *Coefficient of powder.* Some powder placed in density bottle, which is then filled with water. The bottle is warmed, and the mass of water expelled is found. The apparent expansivity of water (relative to the powder and glass) is calculated from "mass expelled/(mass left × temp. rise)", and the powder expansivity is then found by subtracting this from the apparent expansivity of water relative to the glass, found from another experiment without the powder.

Variation of density of water. Water has a maximum density at about 4°C. This is shown by Hope's experiment, and by Joule and Playfair's experiment.

General points. (1) *Correction to barometric height.* The height h_0 of a mercury column at 0°C exerting the same pressure as that due to a height h_t at t°C is given by

$$h_0\rho_0 g = h_t\rho_t g,$$

or $$h_0 = h_t\frac{\rho_t}{\rho_0} = \frac{h_t}{1+\gamma t} = h_t(1-\gamma t), \text{ by binomial expansion.}$$

If the observed height is 758 mm (h_t') on a scale of linear expansivity α accurate at 0°C, then $h_t = 758(1+\alpha t)$. Thus generally,

$$h_0 = h_t'(1+\alpha t)(1-\gamma t) = h_t'[1-(\gamma-\alpha)t].$$

(2) When the temperature of a thermometer is raised from 0°C to 100°C, the mercury in the capillary part increases by an amount $\gamma v_0 t$, where γ is the *apparent* expansivity of mercury, v_0 is the volume of mercury in the bulb to the 0°C level, and $t = 100$°C.

(3) If a 1-litre flask contains mercury, the volume of air above the mercury will remain unaltered for all temperatures when the increase in volume of the mercury = the increase in volume of the flask. Thus $\gamma vt = g\times 1$ litre $\times\, t$, where γ is the true coefficient of mercury, v is the volume of mercury, t is the temperature rise, and g is the cubic expansivity of the container.

(4) The mass m of liquid filling a flask at t°C is given by $m = V_t\rho_t$, where V_t is the volume of the flask at t°C and ρ_t is the density of the liquid at t°C. The volume $V_t = V_0(1+gt)$; the density $\rho_t = \rho_0/(1+\gamma t)$.

13. HEAT CAPACITIES OF GASES ISOTHERMAL AND ADIABATIC CHANGES

The specific heat capacity of a gas depends on the conditions under which it is heated.

Principal specific heats. *The specific heat capacity of a gas at* *constant volume*, c_V, is the heat required to raise the temperature of 1 kg of the gas by 1°C when the volume is constant (Fig. 13.1*a*).

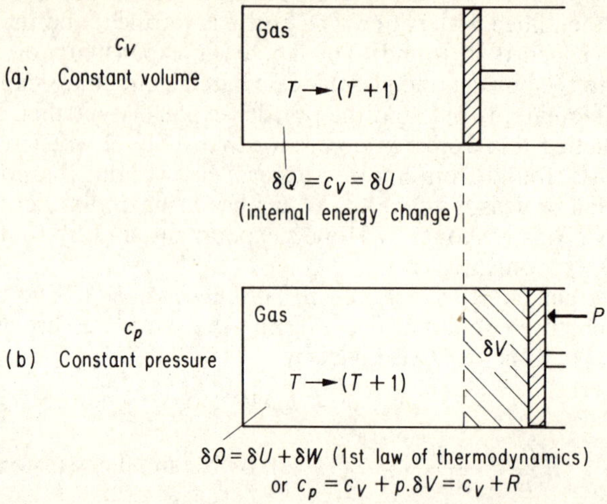

FIG. 13.1. Specific heat capacities of ideal gas.

The specific heat capacity of a gas at constant pressure, c_p, is the heat required to raise the temperature of 1 kg of the gas by 1°C when the pressure is constant (Fig. 13.1*b*).

Unit of c_p or c_V: J kg^{-1} K^{-1}; or, if large, kJ kg^{-1} K^{-1}.

Molar heat capacity. The molar heat capacity at constant volume (C_V) or constant pressure (C_p) is defined as in the case of c_V or c_p respectively, but the mass is now 1 mole.

Unit of C_p or C_V: J mol^{-1} K^{-1} (= kJ kmol^{-1} K^{-1}).

Difference between specific heat capacities. For an ideal gas, one in which there is no attraction between the molecules and in which the molecules have negligible volume, c_V is independent of volume. c_p is greater than c_V by the external work, W, done when the gas expands. Now W = pressure × volume change = $p(V_2 - V_1)$. Since

$$pV_1 = RT \quad \text{and} \quad pV_2 = R(T+1)$$

$$\therefore p(V_2 - V_1) = R.$$

$$\therefore c_p - c_V = R \quad . \quad . \quad . \quad . \quad . \quad . \quad . \quad (1)$$

c_p, c_V, R are in J kg^{-1} K^{-1}.

As an example, suppose c_V = 10·1 kJ kg^{-1} K^{-1} for hydrogen. The volume occupied by 2 kg of hydrogen at 0°C and $1{\cdot}013 \times 10^5$

N m^{-2} pressure is about 22·4 m^3. Then, from $R = pV/T$,

$$R = \frac{1{\cdot}013\times10^5\times22{\cdot}4/2}{273}$$

$$= 4{\cdot}2\times10^3 \text{ J kg}^{-1}\text{ K}^{-1} = 4{\cdot}2 \text{ kJ kg}^{-1}\text{ K}^{-1}$$

$$\therefore c_p = c_V + R = 10{\cdot}1 + 4{\cdot}2 = 14{\cdot}3 \text{ kJ kg}^{-1}\text{ K}^{-1}.$$

Internal energy. (1) c_V is a measure of the *internal energy* per unit mass of a gas. For an ideal gas, the internal energy is independent of volume and depends only on the temperature. For a temperature change from T_1 to T_2, internal energy change per kg $= c_V(T_2-T_1)$.

(2) For a real gas, the Joule-Kelvin effect shows that "internal work" is done when a gas expands. This internal work is the work done against intermolecular attraction. It is additional to the "external work" done against any external pressure. The relation $c_p-c_V = R$ is only true for an *ideal* gas as the internal work is ignored.

(3) When a solid or a liquid is heated it expands slightly; the work done against the external pressure is thus negligible. Hence the two principal specific heat capacities of a solid or a liquid are practically equal. The difference between them is greater for a liquid owing to the greater volume change.

Kinetic theory. Degrees of freedom. The molecules in a *monatomic* gas have only translational kinetic energy. They have a total of three degrees of freedom, in mutually perpendicular directions Ox, Oy, Oz. (Fig. 13.2*a*). Since the translational kinetic energy per unit mass of an ideal gas is $3RT/2$ (p. 113), it follows that the associated energy per degree of freedom $= RT/2$ in this case. Thus the associated energy per molecule $= kT/2$, where k (Boltzmann's constant) $= \mathbf{R}/N_A = 8{\cdot}3/(6\times10^{23}) = 1{\cdot}4\times10^{-23}$ J K^{-1}.

From above, c_V = energy change from T to $(T+1) = 3R/2$, from kinetic theory. Also, $c_p = c_V + R = 5R/2$. Thus $\gamma = c_p/c_V = 5/3 = 1{\cdot}67$, for an ideal gas.

In a *diatomic* gas, molecules have rotational and vibrational kinetic energy in addition to translational kinetic energy (Fig. 13.2*b*). With two degrees of freedom of rotational energy, and assuming the vibrational energy relatively small, then total degrees of freedom = 3 (translational) + 2 (rotational) = 5. Thus $c_V = 5R/2$, from previous. Hence $c_p = c_V + R = 7R/2$. Then $\gamma = c_p/c_V = 7/5 = 1{\cdot}4$. Fig. 13.2*c* shows experimental results of γ for argon (monatomic), hydrogen (diatomic) and air (largely O_2 and N_2), at normal temperatures.

Measurement of c_p. A *constant-flow calorimeter* can be used, with (*i*) platinum resistance thermometers, (*ii*) a coiled tube to slow down the gas flow, (*iii*) a metal gauze to mix the gas after heating, (*iv*) a heater coil over which the gas passes. With the usual notation,

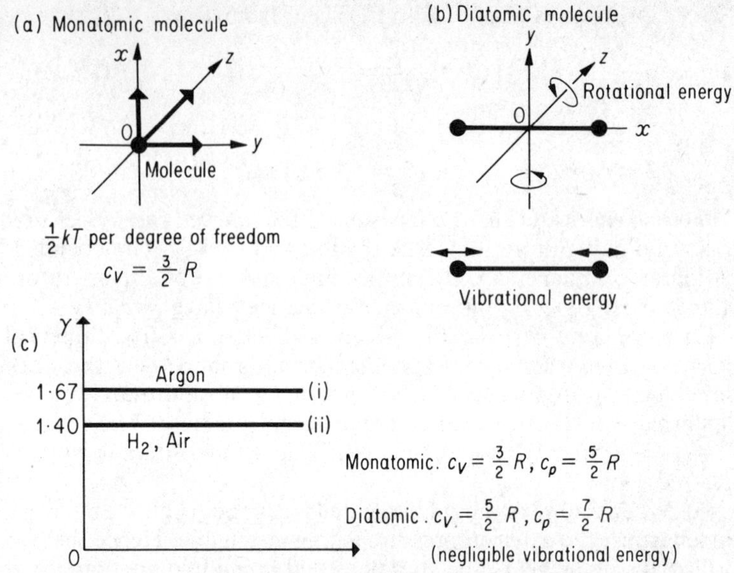

FIG. 13.2. Kinetic energy of molecules and degrees of freedom.

and neglecting heat lost per second by radiation,

$$mc_p(t_2 - t_1) = IV$$

Measurement of c_V. *Joly's differential steam calorimeter used*; one globe evacuated, the other filled with gas. Both globes suspended from scale-pans inside a chamber. Steam passed in, and more steam, m, condenses on globe with gas. Then, with the usual notation, $ml = Mc_V(100 - t)$, where M is the mass of gas.

Reversible change. A gas in a cylinder undergoes a "reversible change" when it is compressed or expanded infinitely slowly by a frictionless piston of negligible weight. No work is then done against friction or the weight of the piston, there are no eddies in the gas, and at any stage of the change the gas can be returned to its original pressure, volume and temperature by reversing exactly the pressure-volume changes. At any stage of the reversible change the condition of the gas is a function only of its pressure, volume and temperature. A reversible change is an ideal change. Irreversible changes occur when friction is present or a sudden expansion occurs.

Reversible isothermal change. This is a reversible change made under conditions of *constant temperature*. A gas expanding or contracting isothermally and reversibly thus obeys Boyle's law, pV = constant. (The magnitude of the constant = mRT.)

The slope of the isothermal p-V curve = $dp/dV = -p/V$.

Work done in reversible isothermal expansion per kg of gas,

$$W = \int_{v_1}^{v_2} p.dV = RT\log_e\left(\frac{V_2}{V_1}\right).$$

Units: W in J when R in J kg^{-1} K^{-1}, T in K.

Note that, for an ideal gas, there is no internal energy change δU in a reversible isothermal change, since there is no change in temperature (Fig. 13.3*a*).

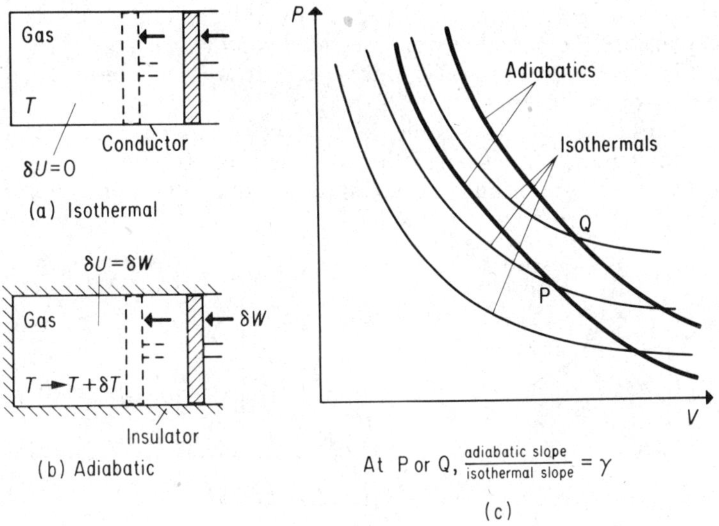

FIG. 13.3. Internal energy in isothermal and adiabatic changes—ideal gas.

Reversible adiabatic change. This is a reversible change made under conditions of *constant heat*, that is, *no heat enters or leaves the gas*. A gas expanding or contracting adiabatically and reversibly obeys the law pV^{γ} = constant, where γ is the ratio of the principal specific heat capacities, c_p/c_V. *To prove*: $\delta Q = \delta U + \delta W = \delta U + p.\delta V$ (First Law of Thermodynamics, or Conservation of Energy) $= 0$ (1); $\delta U = c_V.\delta T$ (2); $c_p - c_V = R$ (3); $p.\delta V + V.\delta p = R.\delta T$ (4), from $pV = RT$. Eliminate δT, and integrate to obtain pV^{γ} = constant. Note that $\delta W = -\delta U$ here.

(1) For temperature calculations, $TV^{\gamma-1}$ = constant, where T is the absolute temperature.

(2) Slope of adiabatic curve $= -\gamma p/V = \gamma \times$ slope of isothermal curve. Adiabatic curves are thus steeper than isothermal curves at their points of intersection, since γ is greater than 1. This is due to the fall in temperature when the volume of a gas increases adiabatically. See Fig. 13.3*c*.

(3) Work done in reversible adiabatic change per kg of gas

$$= \int_{v_1}^{v_2} p.dV = \frac{1}{\gamma-1}[p_2V_2 - p_1V_1]$$

$$= \frac{R}{\gamma-1}(T_2-T_1) = \frac{R}{(c_p/c_V)-1}(T_2-T_1)$$

$$= \frac{Rc_V}{c_p-c_V}(T_2-T_1) = c_V(T_2-T_1)$$

Note that work done adiabatically by 1 kg of an ideal gas on expansion = external work = *decrease in internal energy of gas* = $c_V(T_1-T_2) = -c_V(T_2-T_1)$, since c_V is the change in internal energy per kg of an ideal gas per degC (p.122).

Work done. The work done, W, by a gas, expanding isothermally or adiabatically or in any other way, is proportional to the *area* between the graph and the volume-axis, since this represents the

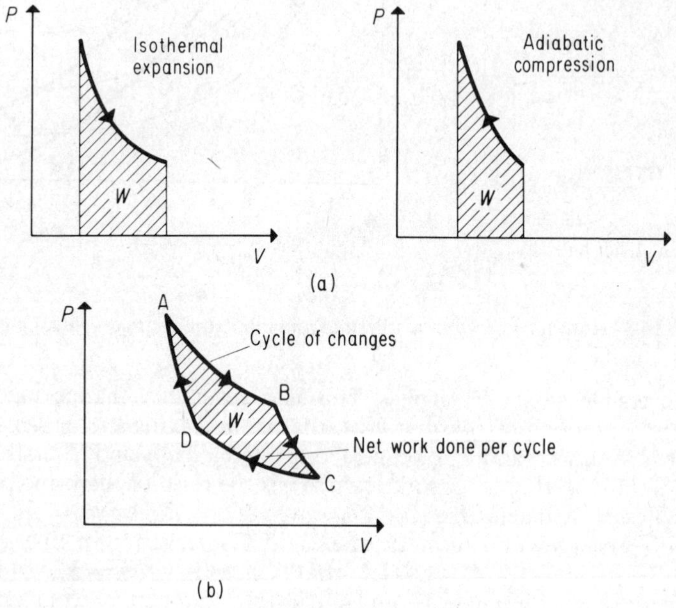

FIG. 13.4. Work done by or on gas.

integral $p.dV$ (Fig. 13.4*a*). When the gas is taken through a complete *cycle* of changes such as ABCDA, the net work done at the end of the cycle is proportional to the area enclosed (Fig. 13.4*b*).

Ratio of principal specific heats. γ can be measured either by (*i*) Clement and Desormes method, or (*ii*) a Kundt's tube experiment.

In (*i*), a manometer with light oil is connected to a large vessel with tap containing the gas, and the pressure is originally slightly above normal (h_1). On opening and closing the tap the manometer finally settles at h_2 above normal. Then $\gamma = h_1/(h_1 - h_2)$.

In (*ii*), the velocity v in the gas is determined by using air as a standard in a Kundt's tube experiment; γ is then calculated from $v = \sqrt{\gamma p/\rho}$, where p is the pressure and ρ is the density of the gas (see p. 95).

For a monatomic gas, $\gamma = 1{\cdot}67$; for a diatomic gas, $\gamma = 1{\cdot}4$. For higher atomicity γ is less, for example 1·29 (p. 123).

First Law of Thermodynamics. This is the Principle of Conservation of Energy. The heat (δQ) supplied to a system is equal to the gain in internal energy (δU) of the system plus the external work (δW) done by the system, or $\delta Q = \delta U + \delta W$. Under reversible conditions, $\delta Q = \delta U + p.\delta V$, where δV is the volume change of a gas at a pressure p.

Second Law of Thermodynamics. (*Clausius.*) It is impossible for any self-acting machine, working in a cyclical process unaided by an external agency, to make heat pass from one body to another at a higher temperature.

(*Kelvin.*) It is impossible, by means of inanimate material agency, to derive mechanical effect from any portion of matter by cooling it below the temperature of the coldest body of its surroundings.

Carnot cycle. The whole of mechanical energy can be converted into heat. The converse is not true. When a gas is taken through a complete cycle of changes, it is impossible to transform *all* the heat obtained from the hot reservoir to mechanical energy—some heat must be rejected to the cold reservoir or sink. The Carnot cycle has the maximum efficiency—it converts more heat to mechanical energy than any other type of cycle.

The Carnot cycle is a reversible series of changes, usually isothermal and adiabatic, in which the working substance is brought back to its initial state.

Stages. (1) Isothermal expansion of gas—heat Q_1 is absorbed at constant temperature, T_1 K, from a hot reservoir. (2) Adiabatic expansion—no heat enters or leaves and the temperature falls. (3) Isothermal contraction—heat Q_2 is given up at constant temperature, T_2 K, to a cold reservoir or sink. (4) Adiabatic compression—no heat enters or leaves and the temperature rises to its initial temperature, and the gas to its original pressure, volume and temperature. Gas laws $pV = RT$ and $pV^\gamma = k$ apply.

Work done during cycle $= Q_1 - Q_2$.

From gas laws, efficiency of cycle =

$$\frac{Q_1 - Q_2}{Q_1} = \frac{T_1 - T_2}{T_1} = 1 - \frac{T_2}{T_1}.$$

VAPOURS

Unsaturated vapour. This is a vapour which is not in equilibrium with its own liquid. It obeys the gas laws approximately. It does not obey the gas laws when near to saturation (Fig. 13.5*a*).

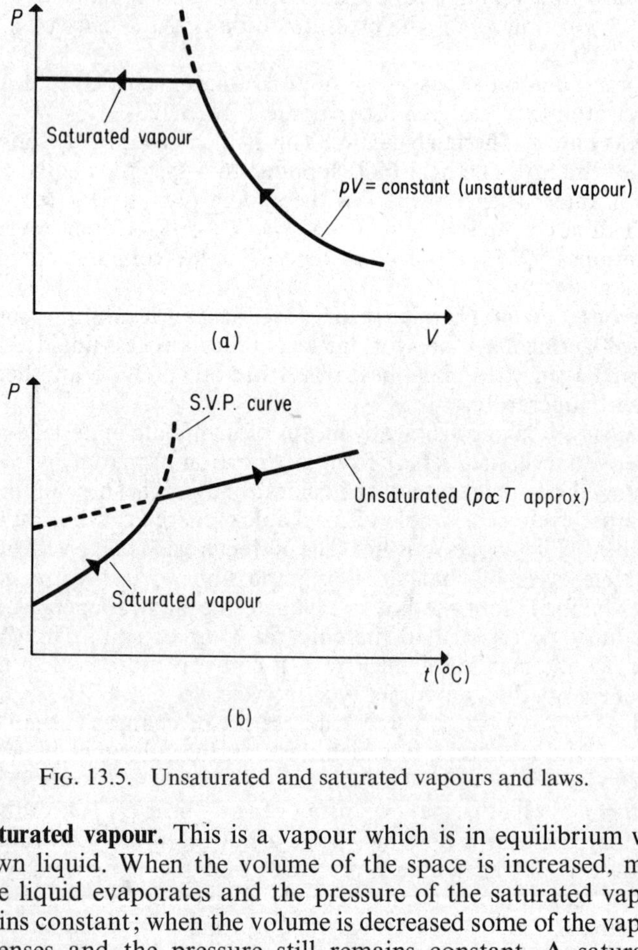

FIG. 13.5. Unsaturated and saturated vapours and laws.

Saturated vapour. This is a vapour which is in equilibrium with its own liquid. When the volume of the space is increased, more of the liquid evaporates and the pressure of the saturated vapour remains constant; when the volume is decreased some of the vapour condenses and the pressure still remains constant. A saturated vapour has thus a maximum density at a given temperature. *The saturation vapour pressure* (*s.v.p.*) *depends only on the temperature of the liquid*, and increases as the temperature increases. (Fig. 13.5*b*). The s.v.p. is independent of the volume. (Fig. 13.5*a*).

A liquid boils at a temperature such that the s.v.p. at that temperature = the external (atmospheric) pressure. The s.v.p. of water-vapour at 100°C is 760 mm mercury.

Variation of S.V.P. of water with temperature. *Static method.* Two barometer tubes, one with vacuum, other with water above mercury. Difference in mercury heights, h=s.v.p. of water at temperature concerned, allowing for errors due to weight of water and density variation of mercury with temperature. Jacket round upper part of barometer tubes for varying the temperature (Fig. 13.6*a*).

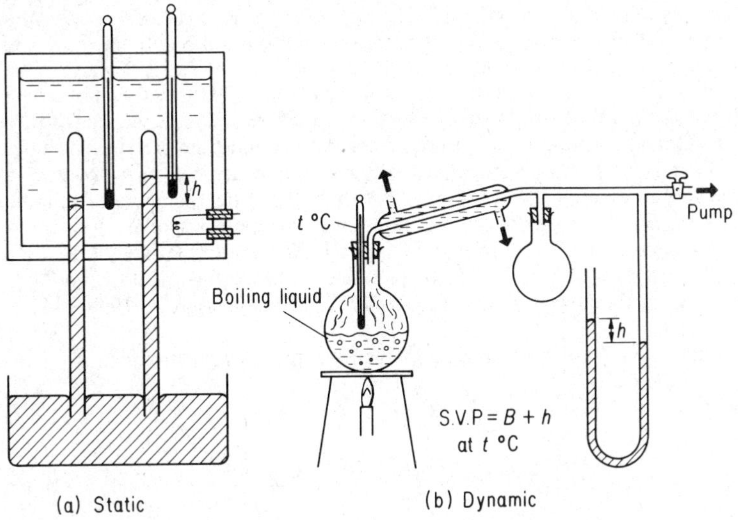

FIG. 13.6. Variation of s.v.p. with temperature.

Dynamic method. Apparatus consists of (*i*) flask with water and thermometer, (*ii*) a condenser arrangement, (*iii*) a mercury manometer, (*iv*) a flask or bulb to allow for sudden pressure fluctuations, (*v*) a pump to vary the pressure. The pressure is altered first, then the water is boiled and the boiling-point observed. The s.v.p. of water = the external pressure, which is $(B \pm h)$ mm of mercury, where B is the barometric pressure and h is the difference in levels of the mercury manometer (Fig. 13.6*b*).

Dalton's law of partial pressures. In a mixture of gases the total pressure is the sum of the pressures of the "parts" (gases), each gas occupying a volume equal to that of the mixture.

Suppose the total pressure in a closed vessel containing air and a little water at 12°C is 900 mmHg. Then $(900-h)$ mm is the pressure of the air if h mm is the s.v.p. of water at 12°C. The volume of the air (or water vapour) = the volume of the vessel. In calculations on mixtures of air and vapour, the rule is to apply the gas laws to the *air* alone; the mass of vapour present may alter when the temperature changes and gas laws apply only to a fixed mass of gas.

EXAMPLE

Compare the properties of saturated and unsaturated vapours. By means of diagrams show how the pressure of (*a*) a gas, (*b*) a vapour vary with change (*i*) of volume at constant temperature, and (*ii*) of temperature at constant volume.

The saturation vapour pressure of ether vapour at 0°C is 185 mm of mercury and at 20°C is 440 mm. The bulb of a constant volume gas thermometer contains dry air and sufficient ether for saturation. If the observed pressure in the bulb is 1000 mm at 20°C, what will it be at 0°C? (*L.*)

First part. The properties of saturated and unsaturated vapours are listed on p. 128. In the case of a gas, the pressure obeys Boyle's law, $pV=$ constant, at constant temperature; at constant volume, the pressure increases uniformly with temperature, according to the relation $p_t = p_0(1+\alpha t)$. In the case of a vapour, cases above and below the critical temperature should be considered; above the critical temperature a vapour acts like a gas, below it condenses as it is compressed (compare Andrews' experiments, p. 131).

Second part. We deal with the pressure of the *air* only at 0°C and 20°C. At 0°C, the air pressure $= p$ say; at 20°C the air-pressure $= 1000-440 =$ 560 mm.

Since $p \propto T$ at constant volume, then, for the fixed mass of air,

$$\frac{p}{560} = \frac{273}{293}$$

$$\therefore p = \frac{273}{293} \times 560 = 522$$

$\therefore$ total observed bulb pressure $= 522+185 = 707$ mm.

Mass of moist air. Suppose the volume of air and water-vapour is 2000 cm^3 at 770 mm mercury pressure, the temperature is 15°C, the dew-point is 6°C, the s.v.p. of water at 15°C, 6°C = 18 mm, 12 mmHg respectively, the density of air at s.t.p. is 1·290 kg m^{-3} and the density of water-vapour is $\frac{5}{8}$th that of air.

We first find the mass of air. Its volume is 2000 cm^3, its temperature is 15°C, and its pressure = (770 − s.v.p. at dew-point) = $770-12=758$ mm. The volume at s.t.p. $= 2000 \times \frac{758}{760} \times \frac{273}{288}$ cm^3; and since 1 m^3 (10^6 cm^3) at s.t.p. weighs 1·290 kg, the mass of air can be calculated.

We now find the mass of water-vapour. Its volume is 2000 cm^3, the same as that of the air, its pressure is 12 mm mercury (not 18 mm), and its temperature is 15°C.

$$\therefore \quad \text{volume at s.t.p.} = 2000 \times \frac{12}{760} \times \frac{273}{288} \text{ cm}^3.$$

Knowing that 1 m^3 (10^6 cm^3) of water-vapour weighs $(\frac{5}{8}) \times 1{\cdot}290$ kg, the volume of water-vapour can now be found. The total mass of the moist air = mass of air + mass of water-vapour.

HYGROMETRY

Relative humidity (R.H.). This is defined as follows:

$$\text{R.H.} = \frac{\text{mass of water-vapour in a given volume of air}}{\text{mass of water-vapour to saturate that volume}} \times 100\,\%.$$

Applying the gas law $pV = mRT$ to the water-vapour, we derive

$$(a)\ \text{R.H.} = \frac{\text{pressure of water-vapour at room temperature}}{\text{s.v.p. of water at room temperature}} \times 100\,\%,$$

$$(b)\ \text{R.H.} = \frac{\text{s.v.p. of water at the dew-point}}{\text{s.v.p. of water at room temperature}} \times 100\,\%.$$

The formula (*b*) is used in practical measurement of relative humidity; it is *not* a definition of relative humidity.

Dew-point. This is the temperature at which the air just becomes saturated with water-vapour. Vapour pressure at room temperature = s.v.p. at dew-point.

Measurement of relative humidity. *Accurate method*: *Regnault (dew-point) hygrometer*. This consists of two test-tubes with highly-polished silvered ends; one test-tube contains ether, and air is bubbled through the ether. Temperature at which moisture is deposited is noted; air-flow then stopped, and temperature noted at which moisture disappears. If t_1°C is the dew-point, and t_2°C the air temperature, then

$$\text{R.H.} = \frac{\text{s.v.p. at dew-point, } t_1\,^\circ\text{C}}{\text{s.v.p. at air-temperature, } t_2\,^\circ\text{C}} \times 100\,\%.$$

The s.v.p. values are obtained from tables; the air-temperature is obtained from a thermometer in the other test-tube. The lower end of this tube is compared continually with the end of the test-tube containing ether, to assist in recognizing when moisture is first obtained.

LIQUEFACTION OF GASES

Andrews' experiments on isothermals of carbon dioxide. Carbon dioxide contained in a capillary tube X by a mercury pellet, the open end of the tube being in a liquid chamber. Air contained in another capillary tube beside X in the same chamber. The pressure p on CO_2 was altered by screwing in a plunger; and assuming air obeyed Boyle's law, p was deduced from observations on the volume of air.

Measurement of $p-V$ values of CO_2 were made at constant temperatures.

Results. Plotting p vertically against V horizontally: (1) Below 31·1°C, liquid only, liquid-vapour, and vapour only are obtained

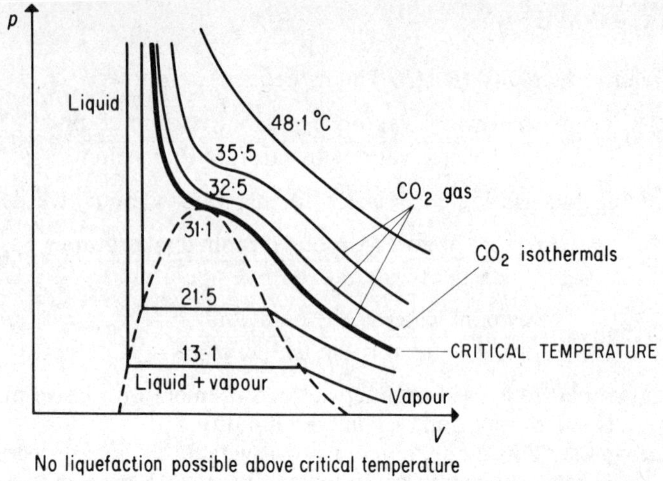

FIG. 13.7. Andrews' experiment on critical temperature.

(Fig. 13.7). The graph is a curve for vapour only stage, a straight horizontal line for the liquid-vapour stage, and a very steep (vertical) line for the liquid only stage. The length of the horizontal line decreases as the temperature increases to 31·1°C.

(2) Above 31·1°C, only gas is obtained. The curve is S-shaped when a few degrees above 31·1°C, but at higher temperatures such as 48·1°C it becomes a smooth curve, as shown in Fig. 13.7.

Deductions. Above 31°C (approx.) CO_2 cannot be liquefied, however great the pressure. This temperature is called the **critical temperature** of the gas.

The **critical pressure** is the pressure at the critical point. The **critical volume** is the volume of unit mass of a substance at the critical temperature and pressure.

A **gas** may be defined as a substance in the gaseous state above its critical temperature.

A **vapour** may be defined as a substance in the gaseous state below its critical temperature.

Joule-Kelvin effect. This is the temperature change due to intermolecular attraction when a gas expands. In most gases, cooling occurs at normal temperatures. In hydrogen and helium, however, the temperature rises. These gases must be cooled below their *inversion temperature* for cooling to occur (see *Liquefaction*, p. 133).

Methods for liquefaction of gases. (1) *Cascade process*. Compression and evaporation utilized in several stages. Methyl chloride evaporates at −24°C, ethylene at −150°C. Oxygen (critical temperature −118°C) then liquefied by compression. Air also liquefied by using the oxygen.

(2) *Linde's method for air.* Joule-Kelvin effect utilized. Air issues through a fine hole, becomes colder on expansion, and flows upward round a tube X through which more air passes. (Fig. 13.8). The air in X is thus progressively cooled, and the critical temperature of air, $-190°C$, eventually passed.

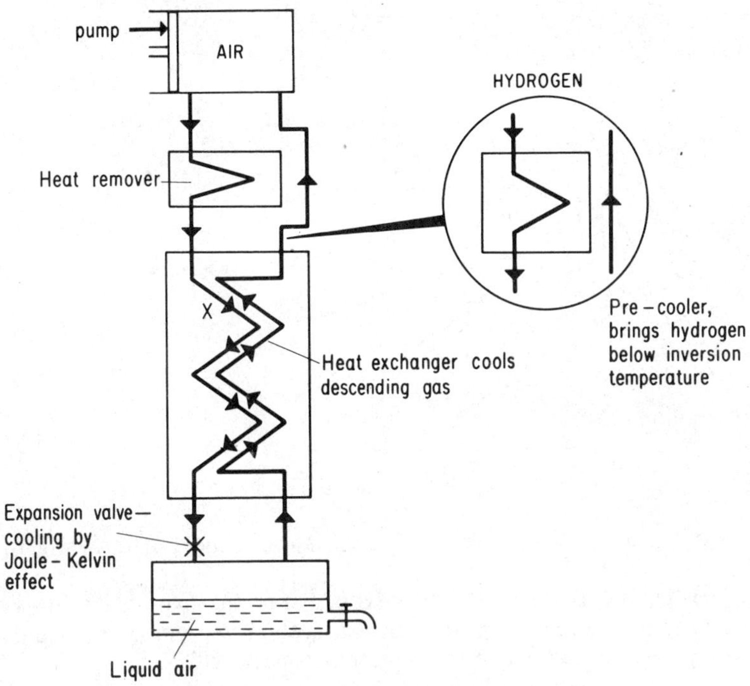

FIG. 13.8. Liquefaction of gases—Linde method.

(3) *Hydrogen.* The Joule-Kelvin effect is positive. The gas is hence cooled below its inversion temperature (p. 132) and then allowed to expand through a fine hole. (Fig. 13.8).

14. THERMAL CONDUCTION AND RADIATION

CONDUCTION

Fundamental formulae. Temperature gradient, g, = temperature difference/distance = $(\theta_1 - \theta_2)/x$. Generally, heat flow per second is given by

$$\frac{Q}{t} = kA\frac{\theta_1 - \theta_2}{x} = kAg \quad . \quad . \quad . \quad . \quad . \quad (1)$$

where Q is the quantity of heat passing through an area A, and k is the thermal conductivity of the material. (Fig. 14.1*a*).

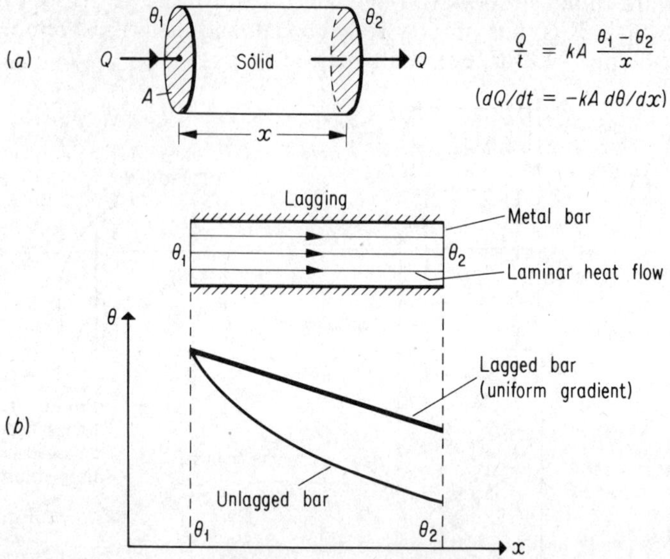

FIG. 14.1 Conduction of heat.

Units: k in W m^{-1} K^{-1}; Q in J; t in seconds; A in m^2; g in K m^{-1} (°C m^{-1}).

Definition of thermal conductivity. From the expression for k, the thermal conductivity is "heat per second per unit area per unit temperature gradient in the direction concerned".

EXAMPLE

The temperature of the air outside a window is −5°C. The average temperature of the room is 15°C. If the glass thickness is 4 mm, the area is 0·5 metre^2, and the thermal conductivity of glass is 1·2 W m^{-1} K^{-1}, calculate the heat lost through the window per minute.

$$g = \frac{15-(-5)}{4\times 10^{-3}} = 5\times 10^{3}\ \text{K m}^{-1}$$

$$\therefore\ Q = kAgt = 1{\cdot}2\times 0{\cdot}5\times 5\times 10^{3}\times 60 = 18\times 10^{4}\ \text{J}$$

Lagged and unlagged bar. The temperature gradient is constant in a lagged bar since Q per second is constant. Hence the temperature-distance (θ v. x) graph is a straight line (Fig. 14.1*b*).

With an unlagged bar heat is lost from the sides. Q per second thus diminishes in magnitude as we proceed from the hot to the cold end.

Consequently the temperature gradient diminishes. The θ v.x graph is therefore a curve of diminishing slope (Fig. 14.1*b*).

Methods of measuring thermal conductivity

Good conductor. *Searle's method.* Apparatus has (*i*) lagged cylindrical thick bar of metal, (*ii*) arrangement, such as steam or heating coil, for heating one end, (*iii*) two mercury thermometers, about 10 cm apart, for the temperature gradient, g, (*iv*) coiled tube round bar, with water passing slowly through it, for the measurement of Q/t (Fig. 14.2). Steady state is attained, the four thermometers are read, the area A of cross-section is measured, and k found from $Q/t = kAg$.

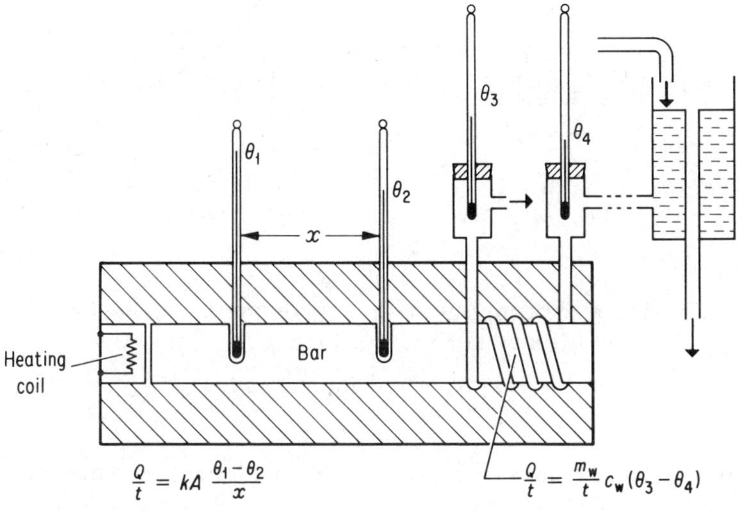

FIG. 14.2. Good conductor—Searle's method.

Note that the lagging ensures linear flow of heat. A thick bar has a high area A so that measurable quantity of heat per second flows.

Bad conductor. *Lees' disc method.* Apparatus has steam chest with bottom made of thick brass plate A, a lower thick brass plate B, and the material, e.g. cardboard, is sandwiched between the plates (Fig. 14.3*a*). Thermometers are placed inside holes drilled in A, B, and temperatures θ_1, θ_2 read in the steady state, e.g. $\theta_1 = 98°C$ and $\theta_2 = 85°C$. To find Q/t, chest is removed, B is warmed gently by a burner until its temperature is a few degrees above 85°C, and with the bad conductor on top of it, B is allowed to cool (Fig.

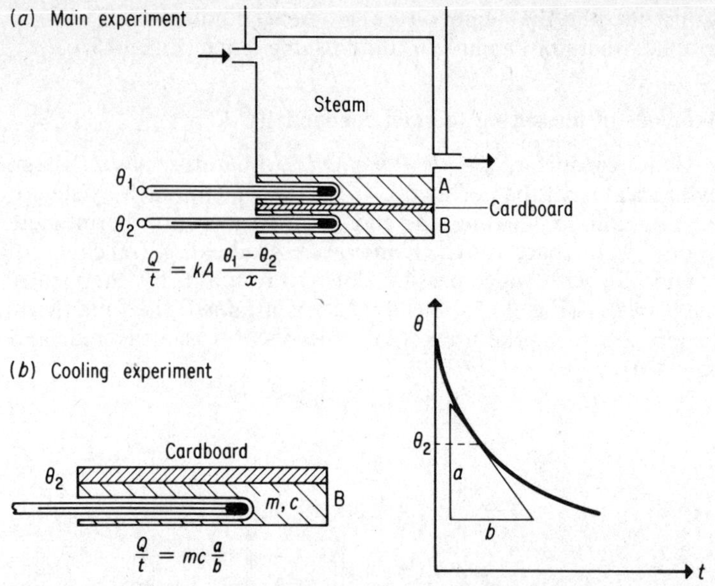

FIG. 14.3. Bad conductor—Lees' disc method.

14.3*b*). From the temperature-time curve, the rate of cooling, a/b, in °C/second is found at 85°C (θ_2). Then

$$Q/t = \text{heat lost by radiation from B at } 85^\circ\text{C}$$
$$= mca/b$$

where m is the mass of B, and c its specific heat. The temperature gradient, g, during conduction $= (\theta_1 - \theta_2)/x$, where x is the thickness of the conductor. k is calculated from the relation $Q/t = kAg$.

Note that the specimen has a large area and is thin so that (*i*) A and g are high and Q/t is then appreciable, (*ii*) little heat is lost by cooling from the curved side and the flow is appreciably linear.

Radial flow. Using the calculus, $Q/t = kAg$ becomes

$$\frac{dQ}{dt} = -kA\frac{d\theta}{dx} \quad . \quad . \quad . \quad . \quad . \quad . \quad (2)$$

θ is the temperature, x is the distance from the hot end. The temperature gradient, $d\theta/dx$, is negative, since θ decreases with increasing x. Thus, from (2), dQ/dt is positive.

With a small electrical heating coil at the centre of a solid metal *sphere*, the heat flow lines are radial. (Fig. 14.4*a*). Consider a section S of the sphere inside it which has a radius r. If the rate of heat supply is W, and conditions become steady, then, from (2), the temperature gradient, $d\theta/dx$, at S is given by

$$\frac{d\theta}{dx} = \frac{d\theta}{dr} = -\frac{W}{k.4\pi r^2}.$$

Thus the temperature gradient decreases according to an inverse-square law.

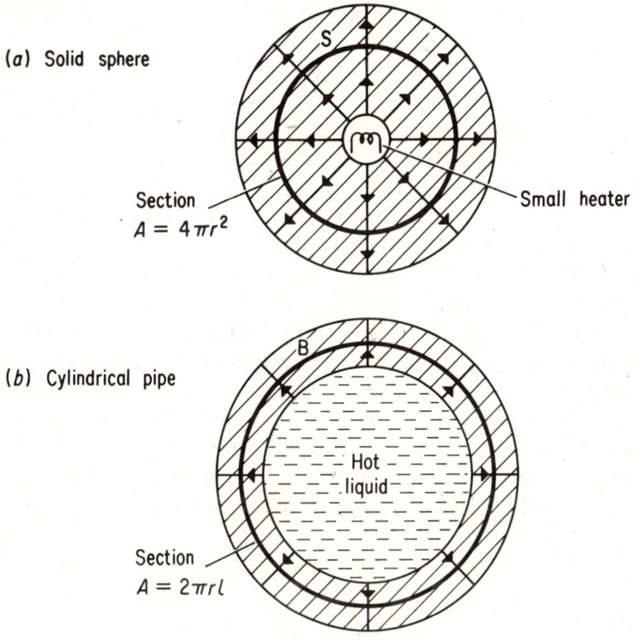

FIG. 14.4. Radial flow—(*a*) sphere (*b*) cylinder.

With a *cylinder* such as a hollow glass tube, or lagging round a pipe, radial heat flow occurs when the inside of the tube or pipe is hot. (Fig. 14.4*b*). This time, $A = 2\pi rl$, where r is the radius of a cylindrical section B in the material and l is the length of the pipe. The temperature gradient, from (2), is proportional to $1/r$.

Thermal conductivity of cylindrical pipe. *Glass in form of a tube.* Pass steam round the outside and water up through the middle. Observe the outlet and inlet water temperatures in the steady state, θ_2, θ_1°C. Then $Q/t = mc_W(\theta_2 - \theta_1)$, where m is the mass of water per second, and temperature gradient

$$g = \left(100 - \frac{\theta_1 + \theta_2}{2}\right)/(r_2 - r_1).$$

The area $A = 2\pi rl$, where l is the length of the tube and r is the average of the inner and outer radii, r_1, r_2 respectively. Apply $Q/t = kAg$ to find k.

Rubber tubing. Pass steam through tubing, then immerse a length l in a calorimeter with water cooled below room temperature θ by ice, and observe the rise in temperature with time for a few degrees above room temperature. From the graph, obtain the rate of rise in temperature, y, at the room temperature. Then $Q/t = (mc_W + C)y$, where m is the mass of water and C is the heat capacity of the calorimeter. The area $A = 2\pi rl$, and the temperature gradient $= (100-\theta)/(r_2-r_1)$, where r_1, r_2 are the internal and external radii of the tube, and r is their average. Apply $Q/t = kAg$ to find k.

Electrical analogy. An electrical potential gradient along a wire produces an electric current, and the current (quantity of electricity per second) is proportional to the potential gradient, from Ohm's law. Similarly, a temperature gradient, g, along a conductor produces a heat flow, and in the steady state $Q/t \propto g$.

When conductors are in series, the electric current in each is the same. Similarly, the same quantity of heat per second flows through conductors in series in the steady state. To calculate the heat flowing per second through (*i*) two or more metals joined in series, with the two extreme temperatures given, or (*ii*) a brick wall with plaster on one side, with the two outer temperatures given, call θ°C the *junction* temperature. Then equate the two values of Q/t for each material, and solve the equation for θ. Substituting the value for θ in either formula for Q/t enables the heat flowing per second to be found. This is illustrated below in Example 2 (*b*).

EXAMPLES

1. In order to maintain a temperature difference of 10°C between the faces of a sheet of cardboard 1·0 mm thick, it is necessary to expend 2000 W m^{-2} of surface, all the energy passing through the cardboard. What is the thermal conductivity of the cardboard?

With the usual notation,

$$Q/t = kA \times \text{temp. gradt.}$$

With $A = 1$ m^2, $Q/t = 2000$ W, temp. gradt. $= 10/(1\times10^{-3})$ K m^{-1}, we have

$$20000 = k \times 1 \times \frac{10}{10^{-3}}$$

$$\therefore k = 0{\cdot}2 \text{ W m}^{-1}\text{ K}^{-1}$$

2. A room has a wall 6 metre by 3 metre made of brick 24 cm thick lined with plaster 1 cm thick with a glass window of area 2 square metre and thickness 5 mm. Assuming that the temperature of the inside surfaces exceeds that of the outside surfaces by 10°C find the heat conducted per second through (*a*) the glass, (*b*) the plaster lined brick. The thermal conductivities of brick, plaster, and glass may be assumed to be respectively 0·56, 0·40, 0·80 W m^{-1} K^{-1}.

With the usual notation,

$$Q/t = kA \times \text{temp. gradt.}$$

(*a*) $$\therefore\ Q/t = 0{\cdot}8 \times 2 \times \frac{10}{5 \times 10^{-3}}$$

$$\therefore\ Q/t = 3200 \text{ W.}$$

(*b*) Since the brick is in series with the plaster, we must first find the temperature of the *junction*. Suppose it is x°C higher than the outside. Then, since the quantity of heat per second through the plaster and brick is the same, we have, equating Q/t for each,

$$0{\cdot}4 \times A \times \frac{10-x}{0{\cdot}01} = 0{\cdot}56 \times A \times \frac{x-0}{0{\cdot}24}$$

$$\therefore\ 10-x = \frac{56x}{40 \times 24}$$

$$\therefore\ x = \frac{1200}{127}\ {}^{\circ}\text{C}$$

$$\therefore\ Q/t \text{ through brick} = 0{\cdot}56 \times 16 \times \frac{1200/127}{0{\cdot}24}$$

$$= 353 \text{ W.}$$

Conduction in liquids and gases. Experiments to measure k complicated by convection in liquids and by convection and radiation in gases. A "Lee's disc" method is arranged for a *liquid* L, which is sandwiched between two copper discs A, B. A glass slab G and a heating coil are arranged above L, the glass being sandwiched between two copper discs. Since heat is applied *above* the liquid, no convection currents are obtained. Knowing the temperatures of A, B, and the heat flowing through L (from the value of k for G), the thermal conductivity of the liquid can be found.

For a *gas*, polished plates X, Y, Z are used. The gas is between Y, Z, and X is placed above Y. A heating coil is in Y, and Z is kept at a constant temperature. By adjusting a heating coil in X, its temperature can be made the same as Y, thereby eliminating heat lost by radiation and conduction from Y in an upward direction.

Linear and radial flow of heat. The fundamental relation for conduction is

$$\frac{dQ}{dt} = -kA\frac{d\theta}{dx}\ (\theta = \text{temperature},\ t = \text{time}).$$

In *linear flow*, if $\frac{dQ}{dt} = c$, a constant (i.e. steady state in a lagged bar), we have, by integration,

$$\theta = -\frac{c}{kA}x + a, \text{ where } a \text{ is a constant.}$$

In *radial flow*, if $\frac{dQ}{dt} = c$, a constant, then, for a sphere, $A = 4\pi x^2$ where x is the radius of a spherical shell.

$$\therefore c = -k.4\pi x^2 \frac{d\theta}{dx}.$$

$$\therefore \int \frac{dx}{x^2} = -\frac{4\pi k}{c}\int d\theta.$$

Integrating between limits r_1, r_2 for x, and θ_1, θ_2 for θ, we obtain

$$\left(\frac{1}{r_1} - \frac{1}{r_2}\right) = \frac{4\pi k}{c}(\theta_1 - \theta_2).$$

For a *cylindrical* shell, $A = 2\pi x l$, where l is the length of the cylinder.

$$\therefore \frac{dQ}{dt} = c = -k.2\pi x l.\frac{d\theta}{dx}$$

$$\therefore \int_{r_1}^{r_2} \frac{dx}{x} = -\frac{2\pi l k}{c}\int_{\theta_1}^{\theta_2} d\theta$$

$$\therefore \log_e\left(\frac{r_2}{r_1}\right) = \frac{2\pi l k}{c}(\theta_1 - \theta_2).$$

For a thin cylinder, $(r_2 - r_1)$ is small. In this case

$$\log_e\left(\frac{r_2}{r_1}\right) = \log_e\left(1 + \frac{r_2 - r_1}{r_1}\right) = (r_2 - r_1)/r_1.$$

See method for thermal conductivity of "glass tubing", p.137.

Growth of ice on pond. With the usual notation, $\delta Q = A\delta x\rho l$ (l = specific latent heat of fusion).

$$\therefore \frac{dQ}{dt} = A\rho l\frac{dx}{dt} = -kA\frac{\theta}{x},$$

where θ is the air temperature immediately above the ice.

Integrating $\therefore t = \frac{\rho l}{2k\theta}x^2$, if $x = 0$ when $t = 0$.

$\therefore x \propto \sqrt{t}$

$\therefore$ the thickness, x, of the ice is proportional to the square root of the time reckoned from zero thickness.

In practice this is an approximate relation.

RADIATION

Good and bad radiators. Best radiator is a dull-black surface, worst radiator is a highly-polished silvered surface. Best absorber of radiation is a dull-black surface, worst absorber is a highly-polished silvered surface. Factory roofs are painted white to keep the inside temperature constant in summer and winter. Ordinary glass absorbs most of the infra-red radiation, and transmits shorter wavelengths.

Experiments to demonstrate: Leslie cube, thermopile or differential air thermometer used.

Nature of radiation. Radiation from a hot body obeys (*i*) laws of reflection, (*ii*) laws of refraction, (*iii*) laws of interference, diffraction, polarization.

To show (*i*), a hot object, a highly-polished plate as reflector, and a thermopile or suitable phototransistor, are required.

To show (*ii*), a rocksalt lens or concave mirrors, a rocksalt prism (glass absorbs most of the radiation), an arc light, and a thermopile are required (see p. 31).

The velocity of radiation is the same as that of light, as the radiation is cut off simultaneously with the light when a total eclipse of the sun occurs. Thus *thermal radiation has the same nature as light*. It is a member of the same family of waves, *electromagnetic waves*, but has a longer wavelength than the visible spectrum. The longest wavelength (red) in the visible spectrum is about 7×10^{-7} m or 0·7 μm. Thermal radiation, called infra-red radiation, has wavelengths of the order of 2 μm.

Prévost's theory of exchanges. With a given system of bodies at different temperatures, an exchange of heat by radiation takes place between the bodies. When the temperatures are all equal the bodies are losing and gaining heat at the same rate, and a state of dynamic equilibrium exists. (Thus an ice-cold body radiates heat.)

Black-body radiation. A black-body or full radiator is one which emits every wavelength, each with the maximum energy possible at the temperature concerned. A black-body radiator is thus a "perfect" or "full" radiator.

The sun is a black-body radiator, but the filament of an electric lamp is not. A black-body radiator in the laboratory can be realized by using a large metal sphere with a small hole in it. Radiation which passes through the hole into the sphere is completely trapped after a few reflections at the walls. The hole is thus a perfect absorber. The hole is hence a black-body absorber (Fig. 14.5*a*). Similarly, at any temperature *T*, the hole emits radiation of all wavelengths, each with the maximum energy possible at that temperature (Fig. 14.5*b*). The hole is a black-body radiator.

Experiment. In a classic experiment, Lummer and Pringsheim used three concentric porcelain tubes, placed a blackened disc

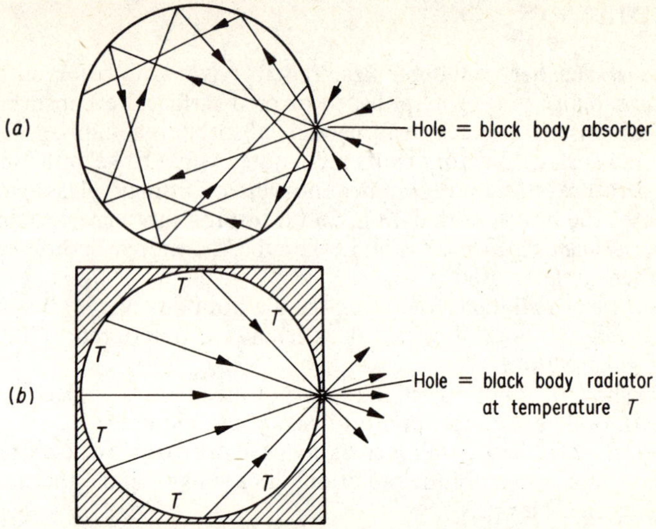

FIG. 14.5. Black body radiator.

inside the innermost tube which was also blackened, and heated the tube electrically. Black-body radiation emerged from the axis of the tube, was dispersed by a fluorspar prism (glass absorbs most of the radiation), and then concentrated by concave mirrors on a *linear thermopile*. This is a narrow strip of platinum, which measures the energy E_λ in a narrow band of wavelengths, λ to $\lambda+\delta\lambda$, in different parts of the spectrum.

Laws of black-body radiation. (1) *Stefan's law*. This states: The total energy radiated per unit area per second by a black-body is proportional to the fourth power of the absolute temperature.

$$\therefore \text{ Energy per unit area per second} = \sigma T^4,$$

where σ is Stefan's constant.

Units: Energy in joule, unit area in metre^2. Since 1 joule per second = 1 watt (W), $\sigma = 5{\cdot}7 \times 10^{-8}$ W m^{-2} K^{-4}.

Under Stefan's law conditions,

$$\text{rate of cooling of body} \propto (T^4 - T_0{}^4),$$

where T_0 is the absolute temperature of the surroundings.

(2) *Energy in different wavelengths*. As λ increases from zero, E_λ increases from a low to a maximum value and then decreases to a low value (Fig. 14.6). As the temperature of the black body increases (*i*) the magnitude of E_λ for a particular wavelength increases, (*ii*) the wavelength λ corresponding to the maximum energy decreases according to the law $\lambda \propto 1/T$, where T is the absolute temperature (*Wien's law*).

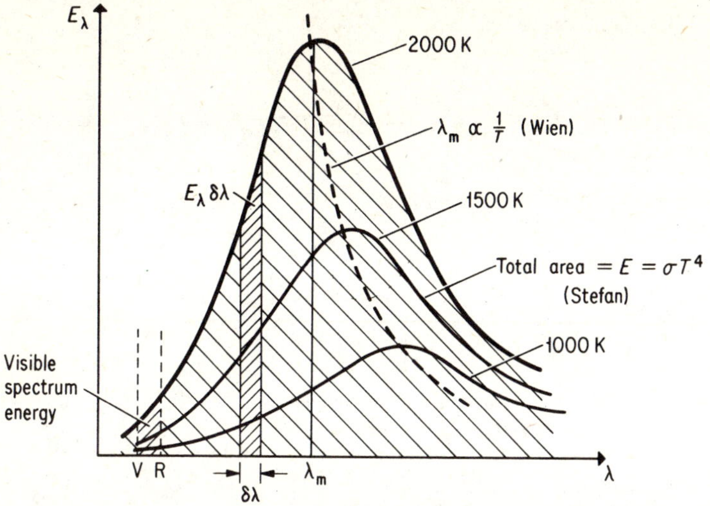

FIG. 14.6. Black body radiation.

Note that E_λ is proportional to the area of a given *narrow strip* of wavelength $\delta\lambda$ at the wavelength λ in the E_λ v. λ curve in Fig. 14.6. In contrast, the total energy E (which obeys Stefan's law) is proportional to the *total area* between the curve and the λ-axis.

Stefan's law applications. If the sun has a radius r and is a black body radiator, then energy radiated per second, $Q_s = \sigma \times \text{area} \times T^4 = \sigma . 4\pi r^2 . T^4$, where T is the sun's radiative temperature in K. Suppose the earth moves in a circular orbit of radius R round the sun. (Fig. 14.7*a*). Then the energy per second per unit area falling normally on the earth's surface = *solar constant* = $Q_s/4\pi R^2 = \sigma r^2 T^4/R^2$. From measurement of the solar constant, T may be found from this relation.

When an electric lamp is switched on (Fig. 14.7*b*), the temperature T of the filament soon becomes constant. This occurs when the rate of energy supplied to the filament from the mains = the rate of energy lost from the surface of the filament (Prévost's law). Thus for a 100 W lamp, ignoring surroundings, T is given by

$$100 = k\,\sigma T^4 . A,$$

where k is a factor (about 0·8) which takes into account that the filament is not a perfect black body radiator, and $A = 2\pi r l$, where r is the radius of the filament and l its length.

Optical pyrometer. This utilizes a narrow band of visible wavelengths, for example red light, and is discussed on p. 100. The instrument is shown in Fig. 14.8*a*, p. 145.

Radiation pyrometer. See page 100 and Fig. 14.8*b* for details. The temperature of the sun's surface can be measured by means of a

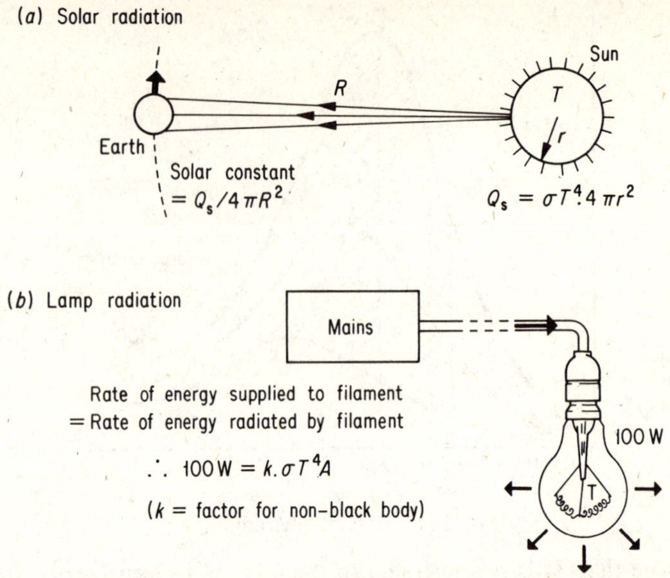

FIG. 14.7 Applications of Stefan's law.

disc rotating steadily in front of the pyrometer. If a sector of angle $\theta°$ is cut out of the disc, the fraction, a, of the radiation incident on the pyrometer is $(360-\theta)/360$. Thus $a\sigma T^4 = \sigma \times 1800^4$, where 1800 K is the supposed reading on the galvanometer and T is the absolute temperature of the sun.

$$\therefore T = \frac{1800}{a^{\frac{1}{4}}} = 1800 \times \left(\frac{360}{360-\theta}\right)^{\frac{1}{4}}.$$

EXAMPLES

1. A hot "black body" of mass 64 g, surface area 16 cm^2 and specific heat capacity 420 J kg^{-1} K^{-1} is allowed to cool down inside an evacuated enclosure surrounded by melting ice. It is found that at 300°C the body cools at the rate of 21°C per minute. Calculate Stefan's constant.

From Stefan's law, the heat radiated per metre^2 per second = σT^4. Hence, taking the enclosure of temperature 273 K into account, heat lost per m^2 per sec at 300°C = $\sigma(573^4-273^4)$

$$\therefore \frac{0{\cdot}064 \times 420 \times 21}{16 \times 10^{-4} \times 60} = \sigma(573^4-273^4)$$

$$\therefore \sigma = \frac{0{\cdot}064 \times 420 \times 21}{16 \times 10^{-4} \times 60 \times (573^4-273^4)}$$

$$= 5{\cdot}7 \times 10^{-8}\ \text{W m}^{-2}\ \text{K}^{-4}$$

2. State Stefan's law of radiation. What is meant by a perfectly black body, and how may it be realized in practice?

Estimate the value of Stefan's constant if the temperature of the filament of a 40-watt tungsten lamp is 2170°C and the effective surface area of the filament is 0·66 cm². You are to assume that the energy radiated is 0·31 of that from a black body in similar conditions and that any effect due to radiation from the glass envelope is negligible. (*L.*)

From Stefan's law for a black-body radiator,

$$\text{energy in joules radiated per m}^2 \text{ per second} = \sigma T^4$$

∴ energy radiated per second from filament

$$= 0{\cdot}31 \times 0{\cdot}66 \times 10^{-4} \times \sigma \times (2170+273)^4 \text{ joule.}$$

Since the temperature is constant, this is equal to the energy supplied per second which is 40 W.

$$\therefore 0{\cdot}31 \times 0{\cdot}66 \times 10^{-4} \times \sigma \times 2443^4 = 40$$

$$\therefore \sigma = \frac{40}{0{\cdot}31 \times 0{\cdot}66 \times 10^{-4} \times 2443^4}$$

$$= 5{\cdot}5 \times 10^{-8} \text{ W m}^{-2} \text{ K}^4$$

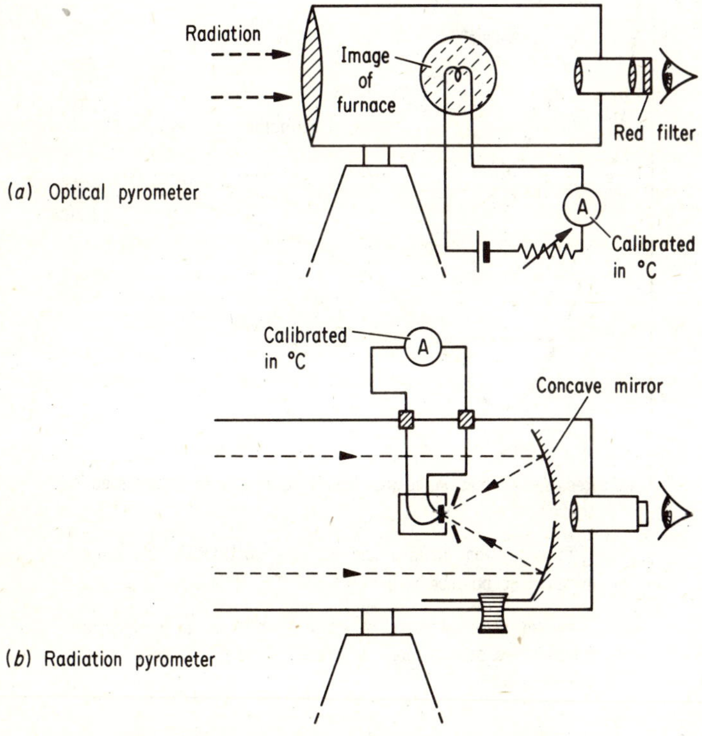

FIG. 14.8. Optical and radiation pyrometers.

Properties of Matter

15. MOLECULAR FORCES. SURFACE TENSION

Intermolecular potential energy and force. Fig. 15.1 illustrates how the potential energy V and force F between two molecules vary with their separation r.

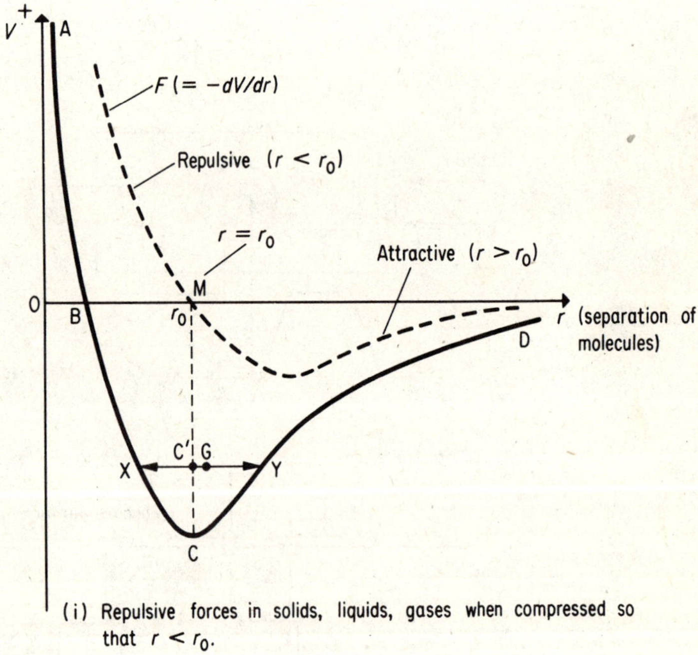

(i) Repulsive forces in solids, liquids, gases when compressed so that $r < r_0$.

(ii) r_0 = separation at minimum V, when attractive and repulsive forces just balance = $3 - 5 \times 10^{-10}$ m (about).

(iii) Minimum value of force at point of inflexion of $V-r$ curve. Beyond this point, force decreases rapidly. At very large r, gaseous state.

FIG. 15.1. Variation in potential energy (V) and force (F) between molecules.

Minimum potential energy, at C, corresponds to the equilibrium separation $r = r_0$ of molecules. At separations greater than r_0, the force is attractive; at separations less than r_0, the force is repulsive. Attractive and repulsive forces counterbalance at $r = r_0$.

On heating, the vibrational energy of the molecules increase. The maximum P.E. may then increase to a level corresponding to XY, and the mean separation r now corresponds to G. Since G is to the right of C′, then $r > r_0$. Thus expansion occurs (see p. 117).

SURFACE TENSION

Experiments demonstrating surface tension. (1) Capillary rise and fall. (2) Spherical shape of a rain drop or mercury droplet or lead shot. (3) Soap bubble. (4) Floating a needle on water. (5) Movement of tiny particles of camphor on water. (6) Lycopodium particles on the surface of water move away from the spot touched by a rod previously dipped into soap solution. (7) A piece of thread, placed gently on a soap film, will be drawn out into a circle when the film enclosed by the thread is pierced.

Surface energy. A molecule A in the bulk of the liquid is surrounded by an equal number of molecules on all sides (Fig. 15.2*a*). The average separation between A and each of the surrounding molecules is such that the attractive and repulsive forces balance. A molecule in the surface, such as C or B, has far more molecules in the liquid

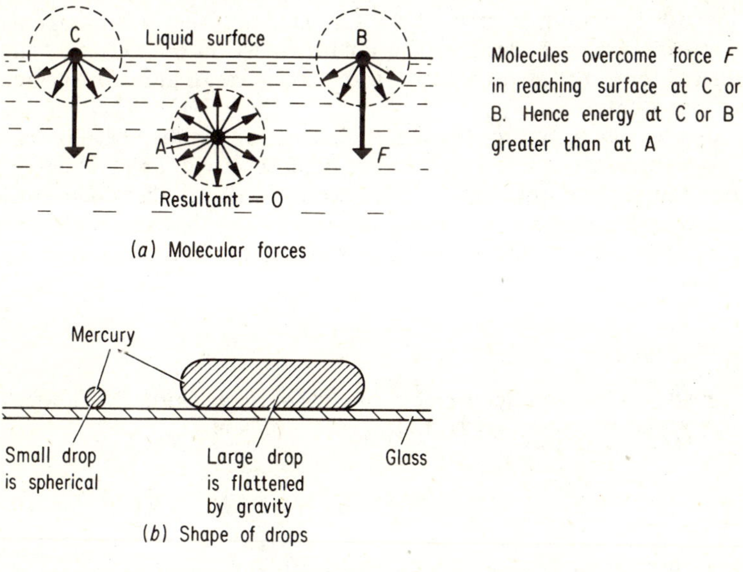

FIG. 15.2. (*a*) Surface tension (*b*) Shapes of drops.

below it than in the vapour above. Thus if all the surface molecules were completely removed, with no liquid temperature change, *energy* would be expended to overcome the resultant attractive force F. Bonds between neighbours have to be broken to bring molecules into the surface from the bulk of the liquid. Hence molecules in the surface have more potential energy than those in the bulk. Surface tension is due to the existence of surface energy.

Shape of drop. Under surface tension forces only, the surface energy is a *minimum*. Thus the surface area is a minimum. A *sphere* is the shape of a given volume of liquid when the area is a minimum. This explains the shape of a falling raindrop.

The area of a sphere $\propto r^2$; the weight of the sphere $\propto$ volume $\propto r^3$. Now area/volume $\propto 1/r$. Thus when r is very small, the surface tension forces predominate over the force due to gravity. On this account a small mercury drop is spherical. When the drop is large, the force due to gravity is relatively more important. The drop becomes flattened at the top, as the centre of gravity is then lowered. See Fig. 15.2*b*.

Coefficient of surface tension (γ). This is defined as *the force per unit length in the surface* acting on one side of, and perpendicular to, a line drawn in the surface.

SI unit: newton per metre, $\mathrm{N\,m^{-1}}$. Dimensions: [force/length] = $\mathrm{MLT^{-2}/L = MT^{-2}}$. At 20°C, $\gamma = 7{\cdot}3 \times 10^{-2}\ \mathrm{N\,m^{-1}}$ for air-water and $45{\cdot}0 \times 10^{-2}\ \mathrm{N\,m^{-1}}$ for air-mercury.

γ *decreases* as the temperature of the liquid rises. Reason: The inward force on molecules at the surface diminishes with temperature rise, as molecules are then farther apart on the average. The energy to overcome the inward force is then less, that is, the surface energy diminishes.

Forces due to surface tension. In calculations, note carefully that γ is a force per unit *length*—do not confuse it with area. Thus if a horizontal ring touches a clean water surface, the downward force on the ring due to surface tension = $2 \times$ *circumference* $\times \gamma$ = $2 \times 2\pi r \times \gamma$, since there are two edges of the ring in contact with the water.

EXAMPLE

Find the downward force F on a thin ring of circumference 10 cm which just touches the surface of water, if $\gamma = 7 \times 10^{-2}\ \mathrm{N\,m^{-1}}$.

$$F = \gamma \times (2 \times \text{circumference})$$
$$= 7 \times 10^{-2} \times 2 \times 0{\cdot}1 = 0{\cdot}014\ \mathrm{N}.$$

In Fig. 15.3*a*, a vertical glass slide just touches a water surface. A rectangular edge is then in contact with the liquid. This has a *perimeter* of $(2a+2b)$, where a is the width and b is the thickness of the slide. The meniscus of the water is tangential to the glass (see *angle of contact*, later). Hence there is a vertically downward force due to surface tension. Its magnitude is $(2a+2b).\gamma$.

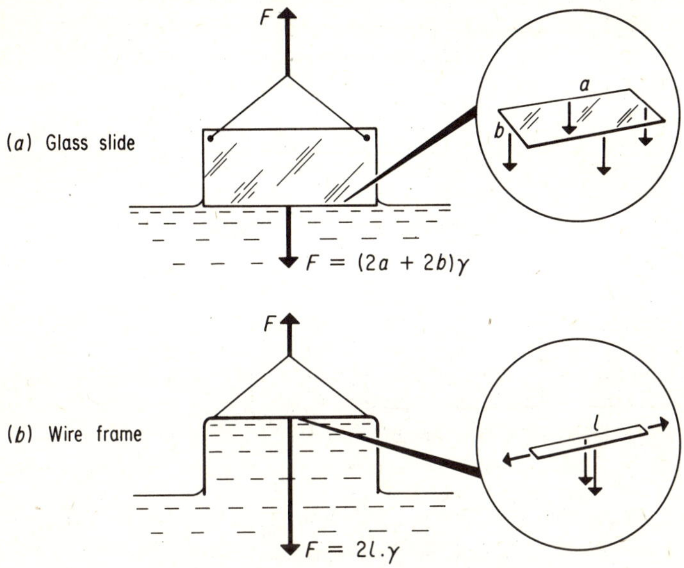

FIG. 15.3. Forces due to surface tension.

In Fig. 15.3*b*, a vertical wire frame contains a soap film. The film has *two* surfaces in contact with air. The horizontal part of the frame, length l, is thus acted on by a vertical force $= 2l.\gamma$, where γ is the surface tension of the soap film.

Angle of contact. The angle of contact, θ, between a liquid and a solid surface in contact with it is defined as the angle between the solid and the tangent to the liquid surface where it meets the solid, *the angle being measured through the liquid.*

For pure water and clean glass the angle is zero, that is, the glass surface is a tangent to the water where the liquid meets it. The angle of contact between alcohol and glass is practically zero. Liquids which have *acute* angles of contact *rise* in a capillary tube. Liquids which have obtuse angles of contact in glass, such as mercury, are depressed in glass capillary tubes.

Measurement of surface tension. Water. *Rise in capillary tube.* $\gamma = rh\rho g/2$, where r is the radius of the capillary C and h is the height of the column of water (Fig. 15.4*a*). The radius is measured by the mercury thread method, or by cutting the tube; h is measured by travelling microscope X, using a pin touching the water level at a point well away from the tube as zero level.

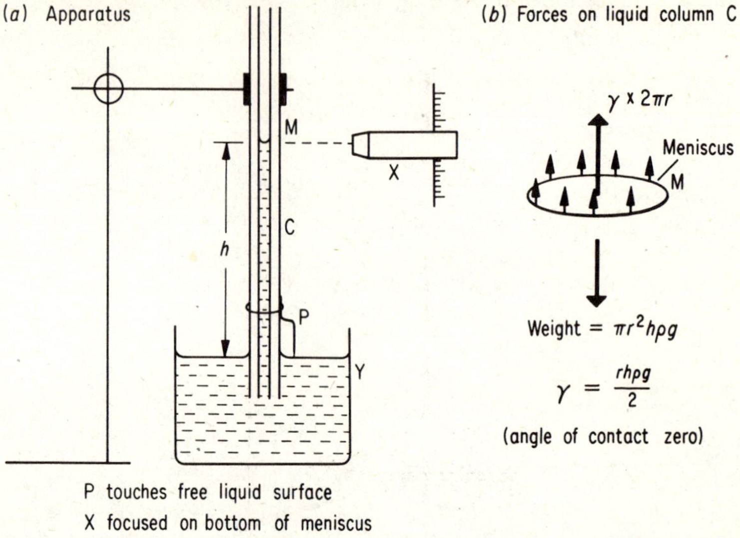

FIG. 15.4. Rise in capillary tube.

Proof of formula: On the *meniscus* M, the glass has an upward force F equal to the downward force due to surface tension. The latter is $2\pi r.\gamma$, assuming zero angle of contact. Hence $F = 2\pi r.\gamma$. On the *column of liquid* supported, the weight acting downward $= F$ acting upwards. Thus $2\pi r.\gamma = mg = \pi r^2 h\rho g$. (Fig. 15.4*b*). Hence, simplifying, $\gamma = rh\rho g/2$.

Units: γ in N m^{-1} when r and h in m, ρ in kg m^{-3}, $g = 9{\cdot}8$ m s^{-2} (10 m s^{-2} approx.).

EXAMPLE

Calculate the height to which water, $\gamma = 7{\cdot}0\times10^{-2}$ N $^{-1}$, will rise in a capillary tube of diameter 0·5 mm.

We have $r = 0{\cdot}25$ mm $= 0{\cdot}25\times10^{-3}$ m, $\rho = 1000$ kg m^{-3}, and suppose $g = 10$ m s^{-2}.

Then

$$h = \frac{2\gamma}{r\rho g} = \frac{2\times7\times10^{-2}}{0{\cdot}25\times10^{-3}\times1000\times10} = 5{\cdot}6\times10^{-2}\text{ m} = 5{\cdot}6\text{ cm}.$$

Microscope slide method. Microscope slide arranged to just touch the water surface. If m is the additional downward force, measured on a sensitive balance, then

$$\gamma = \frac{mg}{2(a+b)},$$

where a, b are the length and breadth respectively of the lower surface of the slide. See Fig. 15.3*a*.

Soap solution. *Film method.* Three-sided wire arranged with soap film in it, and attached to a balance. (See Fig. 15.3*b*.) Then $\gamma = mg/2l$, where l is the length of the upper side of the wire.

Excess pressure in bubbles. *Bubble in water.* The equilibrium of one half B of the bubble gives the following relation between the forces acting on it (Fig. 15.5*a*):

$$2\pi r\gamma + \pi r^2 p_1 = \pi r^2 p_2.$$

$$\therefore p_2 - p_1 = \text{excess pressure} = \frac{2\gamma}{r}.$$

Note that γ multiplies a length ($2\pi r$), whereas p_1 and p_2 each multiplies an area $A(\pi r^2)$.

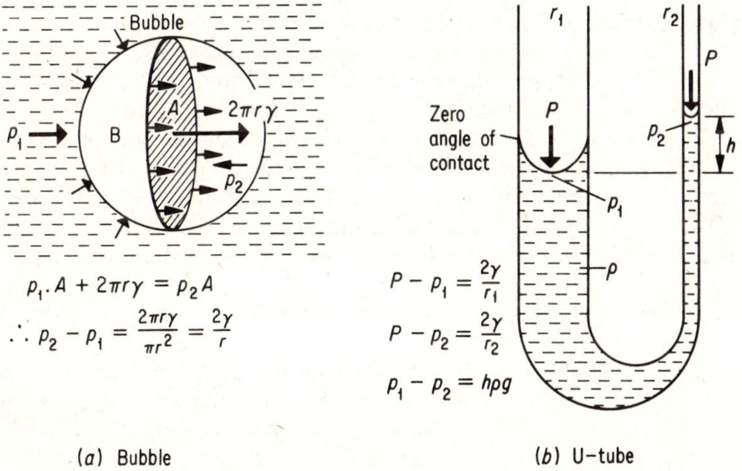

FIG. 15.5. Excess pressure in bubble and curved surfaces.

Soap bubble. Consider the equilibrium of *one half* of the bubble, and remember there are *two* surfaces in contact with air in a bubble.

For equilibrium, then

$$2\times 2\pi r\gamma+\pi r^2 p_1 = \pi r^2 p_2,$$

or $$p_2-p_1 = \text{excess pressure} = \frac{4\gamma}{r}.$$

Meniscus in capillary tube. If the meniscus is tangential to the glass (angle of contact zero), then

$$\text{excess pressure} = \frac{2\gamma}{r}.$$

If the meniscus meets the glass at an angle of contact θ, then

$$\text{excess pressure} = \frac{2\gamma\cos\theta}{r}.$$

Capillary rise and fall by excess pressure method. When a liquid makes an *acute* angle of contact with glass, the liquid in the capillary curves upwards as soon as the latter is dipped into it. The pressure above the meniscus is then greater than below the meniscus by $2\gamma\cos\theta/r$ and the liquid therefore *rises* in the capillary tube to a height h such that $h\rho g = 2\gamma\cos\theta/r$.

When a liquid makes an *obtuse* angle of contact with glass, the liquid in the capillary curves downwards as soon as the latter is dipped into it. The pressure below the meniscus is then greater than that above the meniscus by $2\gamma\cos\theta/r$, where θ is the supplement of the obtuse angle of contact. The liquid therefore *falls* in the tube to a depth h such that $h\rho g = 2\gamma\cos\theta/r$.

Problems in surface tension. (1) In questions on rise or fall in capillary tube, or on difference in levels of a U-tube, the "excess pressure" formula will be found valuable. In the case of the U-tube, (*i*) apply "$p_2-p_1 = 2\gamma\cos\theta/r$" to the meniscus in each side, (*ii*) then use "pressure difference in liquid $= h\rho g$" to calculate h, the difference in levels. See Fig. 15.5*b*.

(2) In a soap-bubble, excess pressure $= 4\gamma/r$; in a bubble in water, excess pressure $= 2\gamma/r$.

(3) The height h to which water rises in a clean capillary tube is given by $h\rho g = 2\gamma/r$, the angle of contact being zero. If a length h_1 of capillary tube less than h is dipped into water, the liquid at the top of the tube then makes an acute angle of contact, θ (it does *not* overflow). Thus $h_1\rho g = 2\gamma\cos\theta/r$. But $h\rho g = 2\gamma/r$. Hence $\cos\theta = h_1/h$.

EXAMPLES

1. Find the pressure inside a bubble of radius 1 mm in water if the external pressure is $1{\cdot}0130\times 10^5$ newton metre^{-2} and γ is 7×10^{-2} N m^{-1}.

$$\text{Excess pressure } p = \frac{2\gamma}{r} = \frac{2\times7\times10^{-2}}{10^{-3}} = 140 \text{ N m}^{-2}$$

$$\therefore \text{ pressure inside} = 1{\cdot}0130\times10^5+140 = 1{\cdot}0144\times10^5 \text{ N m}^{-2}$$

2. The lower end of a vertical tube 2 mm in diameter dips into soap-solution, and on the upper end is a soap bubble 20 mm in diameter. How much is the level of the liquid above that of the surrounding soap-solution? (Surface tension of soap solution $2{\cdot}7\times10^{-2}$ N m^{-1}; the density of soap solution may be assumed to be 1000 kg m^{-3}.

Suppose p is the pressure above the liquid level, or in the bubble, and A is the atmospheric pressure. Then, for the bubble,

$$p-A = \text{excess pressure} = \frac{4\gamma}{r} = \frac{4\times2{\cdot}7\times10^{-2}}{10\times10^{-3}} = 10{\cdot}8 \quad . \quad . \quad \text{(i)}$$

For the liquid meniscus,

$$p-p_1 = \frac{2\gamma}{r} = \frac{2\times2{\cdot}7\times10^{-2}}{10^{-3}} = 54 \quad . \quad . \quad . \quad . \quad . \quad \text{(ii)}$$

where p_1 = pressure in liquid below meniscus.

But $p_1 = A-h\rho g$, by hydrostatics.

$$\therefore p-(A-h\rho g) = 54, \text{ from (ii)}$$

$$\therefore p-A+h\rho g = 54$$

From (i), $\therefore 10{\cdot}8+h\rho g = 54$

$$\therefore h = \frac{43{\cdot}2}{1000\times9{\cdot}8} \text{ metre} = 4{\cdot}4 \text{ mm.}$$

Variation of surface tension. The surface tension of a liquid decreases (*i*) as the temperature increases, (*ii*) as the purity decreases, A vertical film of water cannot be formed on a frame because the surface tension of a pure liquid is constant, and hence there is no excess upward force to support the film. A vertical soap film can be formed on the frame because the soap drains downward to the bottom, thus decreasing the surface tension here. The surface tension force at the top of the film is greater than at the bottom, and the excess upward force can support the film. The variation of the surface tension of water with temperature can be investigated by a bubble method (Jaeger's method) (Fig. 15.6*a*,*b*). Here a bubble, formed slowly in water, L, at the end of A, breaks away when its radius is equal to the radius of A. If the maximum pressure $h\rho g$ is observed in M, then $2\gamma/r = h\rho g - h_1\rho_1 g$.

Surface tension and energy. When the surface area of a film is increased isothermally, the total increase in the surface energy = work done against surface tension forces + heat flowing in from surroundings. The work done against surface tension forces in

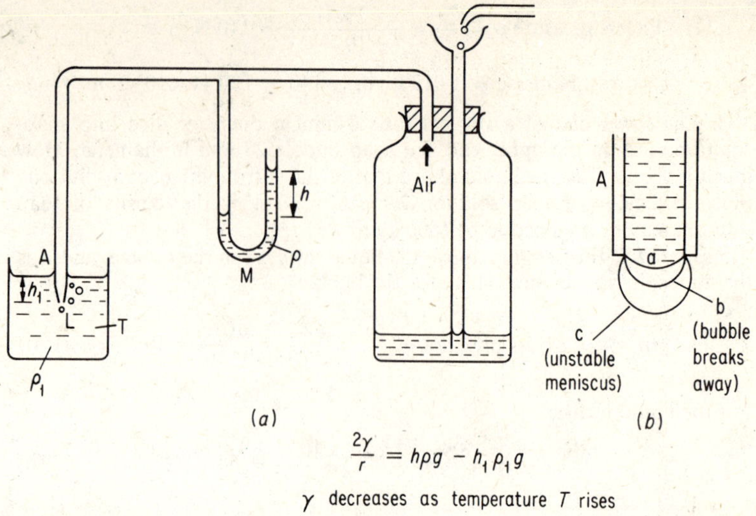

FIG. 15.6. Variation of surface tension with temperature.

enlarging the area by 1 m^2 is called the *free surface energy*, E, and is numerically equal to γ, the surface tension.

In boiling a liquid, energy is needed to change from the liquid to the vapour state, i.e. energy is needed to break the bonds between molecules and their neighbours. The latent heat of vaporization, l, of a liquid hence follows the same variation as E with temperature; each represents energy required to break bonds. Both E and l

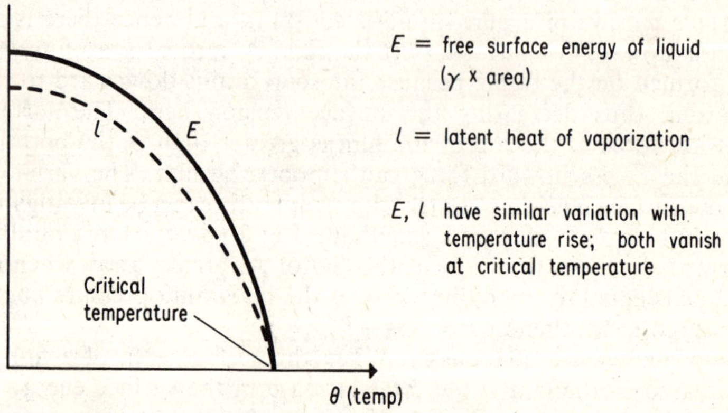

FIG. 15.7. Variation of free surface energy with temperature.

become zero at the critical temperature, above which no liquid state or meniscus is possible. Fig. 15.7.

For a soap-bubble of radius 2 cm, the work done in enlarging the surface isothermally by blowing the bubble $= \gamma \times$ increase in area

$$= \gamma \times 8\pi r^2 = 2{\cdot}7 \times 10^{-2} \times 8\pi \times 0{\cdot}02^2$$

$$= 2{\cdot}7 \times 10^{-4} \text{ joule}$$

if $\gamma = 2{\cdot}7 \times 10^{-2}$ N m^{-1} for soap solution.

When two bubbles of radii r_1, r_2 coalesce under isothermal conditions, the radius r of the bubble formed is given, from the formula for work done, by

$$\gamma \times 8\pi r^2 = \gamma \times 8\pi r_1{}^2 + \gamma \times 8\pi r_2{}^2,$$

or

$$r^2 = r_1{}^2 + r_2{}^2.$$

16. ELASTICITY

Definitions. Elastic. A wire is said to be *elastic* if it returns to its original length when the load on it is removed.

Hooke's law. Provided the elastic limit is not exceeded, the extension of a wire is directly proportional to the tension or load applied.

Elastic limit. The elastic limit is the load beyond which a wire does not obey Hooke's law and the wire does not return to its original length when the load is removed.

Yield point. This is the point just beyond the elastic limit when the extension of the wire increases rapidly as the load is increased, and the wire becomes plastic. A "ductile" material has a yield point but not a "brittle" one.

Young's modulus (E) = tensile stress/tensile strain. The tensile stress is the *force per unit area*; the tensile strain is the *extension/original length* (Fig. 16.1a).

Units: N m^{-2} (same units as "stress" since "strain" is a ratio only). Dimensions: ML^{-1} T^{-2}. For steel, $E = 2{\cdot}1 \times 10^{11}$ N m^{-2}.

Formulæ. If E is Young's modulus, F is the force applied, A is the cross-sectional area of the wire, e is the extension, and l is the original length, then

$$E = \frac{F/A}{e/l}, \quad \ldots \ldots \quad \text{(i)}$$

so that

$$F = EA\frac{e}{l} \quad \ldots \ldots \quad \text{(ii)}$$

Units: F in N when E in N m^{-2}, A in m^2, e and l in m.

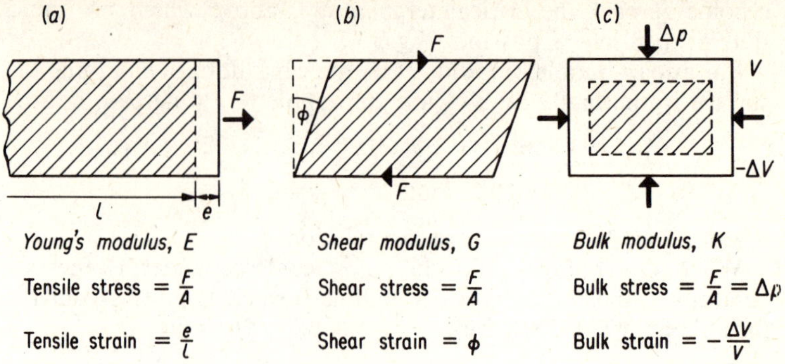

FIG. 16.1. Tensile, shear, bulk stress and strain and modulus.

Experiments. (1) *Young's modulus.* Two long similar wires, P, Q, are suspended from the same beam B; one carries a millimetre scale M, the other a vernier V (Fig. 16.2*a*). P is kept taut by a fixed weight A; loads from 1 to 6 kgf in 1 kgf steps are added to, and then removed

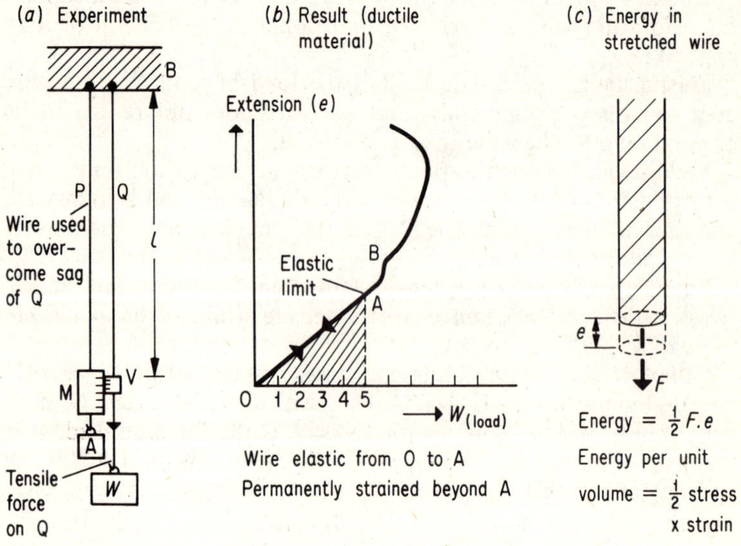

FIG. 16.2. (*a*) (*b*) Young's modulus experiment (*c*) Energy in stretched wire.

from, Q, and the mean corresponding readings *a*, *b*, *c*, *d*, *g*, *f* on the vernier are noted. The average extension *e* for 3 kgf is calculated from $\frac{1}{3}(d-a+g-b+f-c)$. Diameter of wire x measured in several places. Then

$$E = \frac{F}{A} \times \frac{l}{e} = \frac{3 \times 9{\cdot}8 \times l \times 4}{e \times \pi x^2},$$

where l is the length of the wire from the beam to the vernier V.

(2) *Elastic limit.* Increasing loads added in steps of 1 kgf, and extension plotted against load. The elastic limit, A, corresponds to the point just beyond the *limit of proportionality* (end of straight line). (see Fig. 16.2*b*). The yield point is B.

(3) *Verification of Hooke's law.* Plot the extension against the load for moderate load values. A straight line graph passing through the origin is obtained, showing the extension is directly proportional to the load.

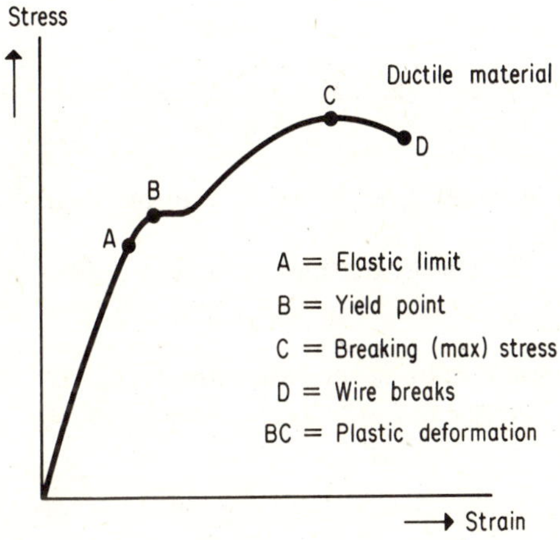

FIG. 16.3. Stress-strain variation for ductile material.

Fig. 16.3 shows the stress-strain variation for a ductile material. Some of the desirable properties of two materials are illustrated in Fig. 16.4*a*, *b*. These are for (*a*) lift cables, (*b*) aeroplane wings.

Note. In stretching a wire by a moderate load, (*i*) the tensile stress is constant at all points of the wire of length l, (*ii*) the extension is proportional to the length, so that the total extension e is the sum of the extensions from the top to the bottom of the wire.

Applications. *Force in wire prevented from contracting.* Since $e = \alpha l t$, where t is the temperature rise and α is the linear expansivity,

$$F = EA\alpha t.$$

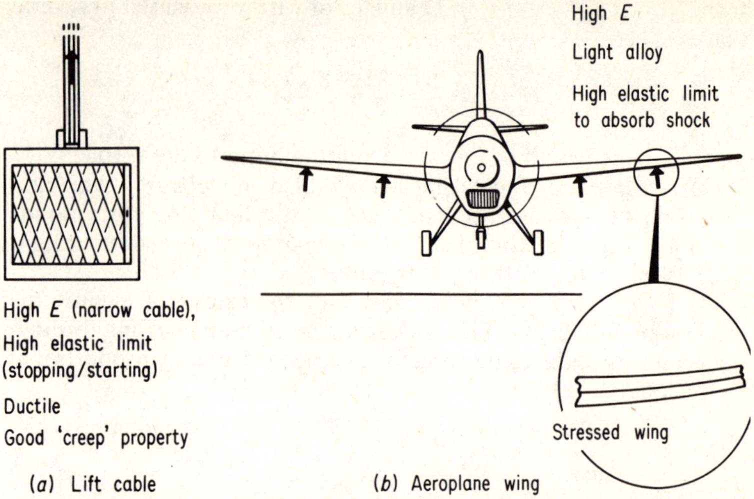

FIG. 16.4. Properties of materials for (*a*) lift cable (*b*) aeroplane wing.

Units: F in N when E in N m^{-2}, A in m^2, α in K^{-1} (°C^{-1}), t in °C.

Energy in wire. The energy per unit volume $= \frac{1}{2}$ stress × strain.

$$\text{Total energy} = \tfrac{1}{2}Fe,$$

where F is the load and e is the extension. (Fig. 16.2*c*).

Units: Total energy in J when F in N and e in m.

EXAMPLES

1. Find the force F required to extend a wire 2 metres long and cross-section area 0·3 mm^2 by 0·5 cm. What is the energy stored in the wire? ($E = 2\times10^{11}$ N m^{-2}.)

(*i*)
$$F = EA\frac{e}{l}$$
$$= 2\times10^{11}\times\frac{0{\cdot}3}{10^6}\times\frac{0{\cdot}5}{100\times2}\text{ N}$$
$$= 150\text{ N} = 15\text{ kgf (approx.)}.$$

(*ii*)
$$\text{Energy stored} = \tfrac{1}{2}Fe = \tfrac{1}{2}\times150\times\frac{0{\cdot}5}{100}$$
$$= 0{\cdot}4\text{ J (approx.)}.$$

2. A vertical steel wire and a parallel brass wire, each 2 metre long and 0·2 mm in diameter, hang from the ceiling and are 50 cm apart. The lower ends are attached to points 50 cm apart on a light horizontal bar. What weight must be hung from the bar to extend both wires by 5 mm, and at what distance from the steel wire must it be attached? [Young's modulii: steel 2×10^{11} N m^{-2}; brass 10^{11} N m^{-2}.]

Since $F = EAe/l$, an extension e of 5 mm requires a force F_1, for steel, given by

$$F_1 = 2 \times 10^{11} \times \pi \times (0{\cdot}1 \times 10^{-3})^2 \times \frac{5}{2000}\ \text{N}$$

$$= 15{\cdot}7\ \text{N}.$$

Similarly, for brass,

$$F_2 = 10^{11} \times \pi \times (0{\cdot}1 \times 10^{-3})^2 \times \frac{5}{2000}\ \text{N}$$

$$= 7{\cdot}9\ \text{N}$$

$\therefore$ total weight required $= 15{\cdot}7 + 7{\cdot}9 = 23{\cdot}6$ N $= 2{\cdot}4$ kgf (approx.).

This is placed $\frac{1}{3} \times 50$ or $16\frac{2}{3}$ cm from the steel wire.

Bulk modulus. The bulk modulus, K, = bulk stress/bulk strain.
Bulk stress = applied force per unit area (= pressure δp).
Bulk strain = volume change/original volume ($= -\delta V/V$). See Fig. 16.1(*c*), p. 156.

$$\therefore K = -\frac{\delta p}{\delta V/V} = -V\frac{dp}{dV}.$$

Units: K in N m^{-2} when δp in N m^{-2}. Fig. 16.5 illustrates how K changes for solids, liquids and gases.

Isothermal bulk modulus of gas. In isothermal changes, pV = constant. By finding dp/dV, it can now be shown that $-V\dfrac{dp}{dV} = p$.

$\therefore$ isothermal bulk modulus $= p$. See Fig. 16.5.

Adiabatic bulk modulus of gas. In adiabatic changes, pV^γ = constant. By differentiation, adiabatic bulk modulus $= -V\dfrac{dp}{dV} = \gamma p$.

The velocity of sound in a gas, $v = \sqrt{\gamma p/\rho}$. The r.m.s. value $\sqrt{\overline{c^2}}$ of the velocity of the molecules of a gas $= \sqrt{3p/\rho}$ from the kinetic theory of gases.

$$\therefore \frac{v}{\sqrt{\overline{c^2}}} = \sqrt{\frac{\gamma}{3}}.$$

Shear modulus or modulus of rigidity, G,
= shear stress/shear strain.
= force per unit area/ angle of twist.
$= F/A \div \phi$. See Fig. 16.1*b*, p. 156.

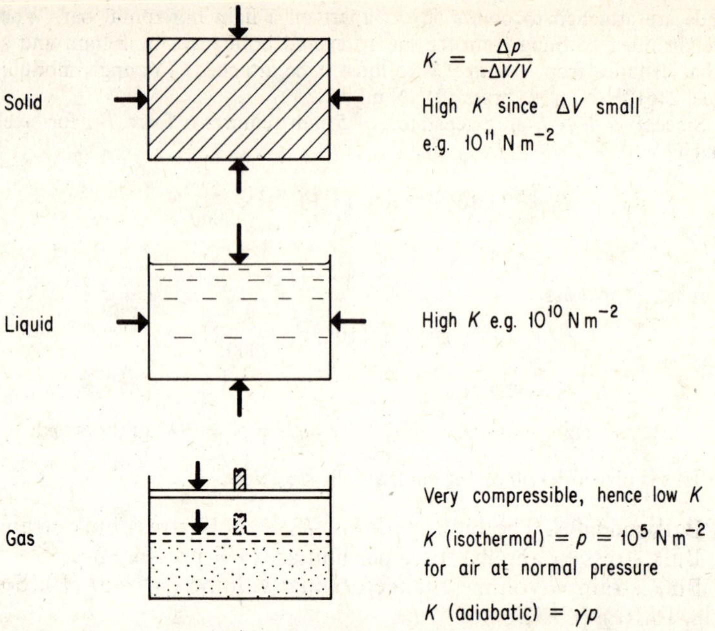

FIG. 16.5. Values of bulk modulus.

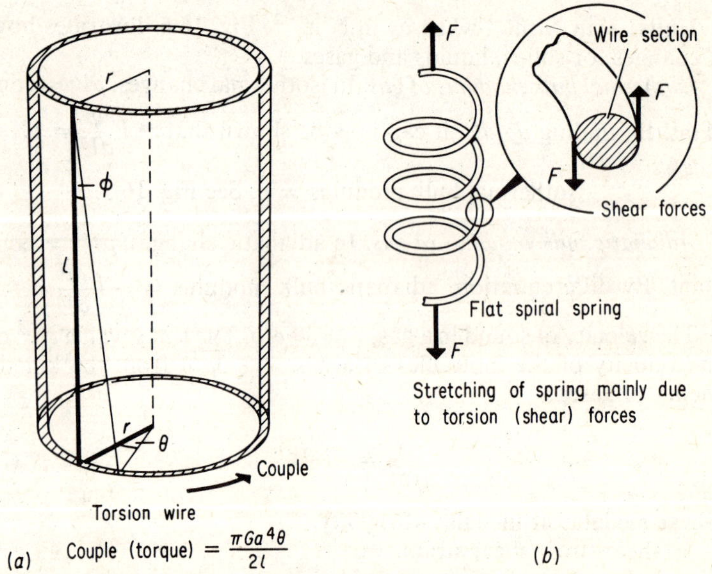

FIG. 16.6. Shear (or torsion) in wire and spring.

When a wire is fixed at one end and twisted through an angle θ at the other end by a couple, the torque C of the couple is given by

$$C = \Sigma F.r = \int_0^a \frac{2\pi G\theta r^3 dr}{l} = \frac{\pi G a^4 \theta}{2l},$$

where a is the radius of the wire, and l is its length. See Fig. 16.6*a*.

The stretching of a flat spiral spring by a load is due mainly to shear forces, F. (Fig. 16.6*b*).

Measurement of G. (1) *Static method.* Apply a couple of torque C at one end by means of a wheel; measure the deflection θ with a pointer and scale for various values of C. Obtain C/θ from a graph, and substitute in $G = \dfrac{2l}{\pi a^4}\left(\dfrac{C}{\theta}\right)$.

(2) *Dynamic method.* Suspend a disc of known moment of inertia I by the wire, and determine the period T of torsional oscillations of the disc. Then, from the mechanics of a rigid body

$$T = 2\pi\sqrt{\frac{I}{c}} = 2\pi\sqrt{\frac{I.2l}{\pi G a^4}},$$

and hence G can be calculated.

Poisson's ratio, σ. This is defined by the ratio

$$\sigma = \frac{\text{fractional lateral contraction}}{\text{fractional longitudinal extension}}$$

$$= \frac{\text{lateral contraction/original diameter}}{\text{longitudinal extension/original length}}$$

Thus $\quad \sigma = -\dfrac{\delta r/r}{\delta l/l}.$

If the volume of the wire is constant when a force is applied, then $V = \pi r^2 l =$ constant. By differentiation, $-\delta r/r = \frac{1}{2}\delta l/l$. Thus Poisson's ratio $= \frac{1}{2}$ in this case. If the wire decreases in volume when stretched, σ is less than 1/2.

17. SOLID FRICTION. VISCOSITY

SOLID FRICTION

Laws of solid friction. (1) The frictional force between two surfaces opposes their relative motion.

(2) The frictional force is independent of the area of contact of the given surfaces when the normal reaction is constant.

(3) The limiting frictional force is proportional to the normal reaction for the case of static friction. The frictional force is pro-

portional to the normal reaction for the case of dynamic friction, and is independent of the relative velocity between the surfaces.

Coefficient of static friction, μ, $= F/R$, where F is the limiting static frictional force and R is the normal reaction.

Method of measuring the static coefficient between two materials: (*i*) block attached by string to scale-pan, and weights added until the block is about to slip. Thus limiting frictional force, F, found. The reaction R = weight of block. Then $\mu = F/R$. Experiment repeated for different weights on block A (see Fig. 17.1*a*).

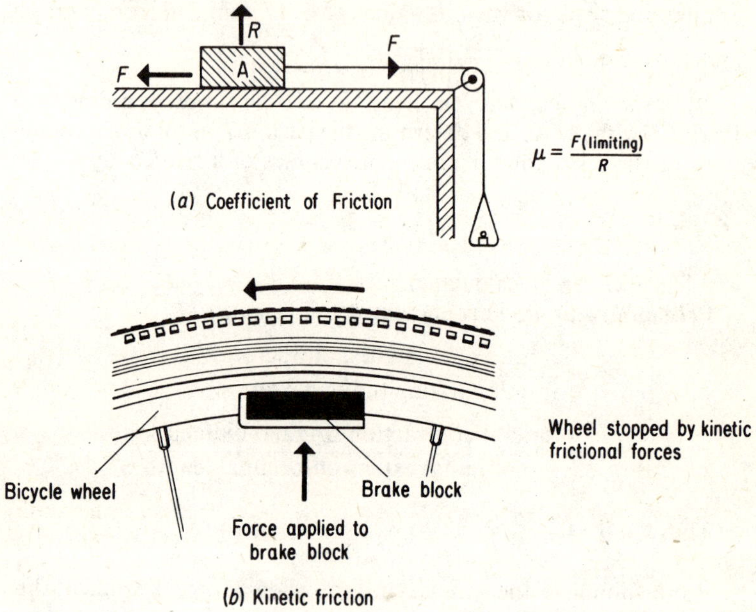

Fig. 17.1. (*a*) Coefficient of static friction (*b*) Kinetic friction.

(*ii*) Block placed on plane, which is gradually tilted until it just slips. Then $\mu = \tan\theta$, where θ is the inclination of the plane to the horizontal.

Coefficient of kinetic (dynamic) friction, $\mu' = F'/R$, where F' is the frictional force when one surface is moving over the other. See Fig. 17.1*b*.

To measure the kinetic coefficient: The block is attached to a scale-pan, and given a slight push each time weights are added to the scale-pan until the block continues to move with a constant velocity. Thus F' is found; the reaction R = weight of block. $\mu' = F'/R$, and is found to be less than the static coefficient.

VISCOSITY

Coefficient of viscosity. The coefficient of viscosity of a liquid, η, is defined as the frictional force acting per unit area per unit velocity gradient.

$$\left[\text{Velocity gradient} = \frac{\text{velocity change}}{\text{distance}} = \left(\frac{dv}{dr}\right)\right].$$

Unit: $N\ s\ m^{-2}$ (SI). ("Poise" in CGS units. To convert values in poise to SI units, divided by 10.)

Dimensions: [velocity gradient, g] = T^{-1}. Dimensions of η = $ML^{-1}T^{-1}$ (*memorise*).

Viscous force. $F = \eta Ag$.

Unit: F in N when η in $N\ s\ m^{-2}$, A in m^2, g in s^{-1}.

EXAMPLES

1. Find the frictional force on a surface of area 100 cm^2 in a region of velocity gradient 0·5 s^{-1} inside a liquid of coefficient of viscosity 2×10^{-3} $N\ s\ m^{-2}$.

$A = 100\times10^{-4}\ m^2$, velocity gradient $= 0{\cdot}5\ s^{-1}$.

$$\therefore F = \eta A \times \text{velocity gradient}$$
$$= 2\times10^{-3}\times(100\times10^{-4})\times0{\cdot}5$$
$$= 10^{-5}\ N.$$

2. On the assumption that the flow of liquid in a tube becomes turbulent at some critical velocity V which depends only on the viscosity η, the density of the liquid ρ, and the radius of the tube r, find how V varies with these other quantities.

$$\text{Let } V = k.\eta^x, \rho^y, r^z.$$

The dimensions of η are $ML^{-1}T^{-1}$; the dimensions of ρ are ML^{-3}; the dimension of r is L; the dimensions of V are LT^{-1}.

$$\therefore LT^{-1} = (ML^{-1}T^{-1})^x(ML^{-3})^y(L)^z.$$

Equating indices on both sides, we have

$$0 = x+y;\quad 1 = -x-3y+z;\quad -1 = -x.$$

Solving, $\therefore x = 1,\quad y = -1,\quad z = -1$

$$\therefore V = \frac{k\eta}{\rho r}, \text{ where } k \text{ is a number.}$$

Flow of liquid through pipe. We distinguish between *laminar* (*orderly*) *flow* and *turbulent flow* of liquid. The formula for laminar flow was first obtained by Poiseuille, and states:

$$Volume\ per\ second = \frac{\pi p a^4}{8\eta l},$$

where p is the difference in pressure between the ends of a pipe of radia a and length l, and η is the coefficient of viscosity of the liquid flowing along it.

Units: Volume per second in $m^3\ s^{-1}$ when p in $N\ m^{-2}$, a in m, η in $N\ s\ m^{-2}$, l in m.

Proof of formula by dimensions. Let the volume per second $= k\eta^x a^y g^z$, where $g = p/l$ = pressure gradient along pipe. Using dimensions,

$$L^3T^{-1} \equiv (ML^{-1}T^{-1})^x L^y (ML^{-2}T^{-2})^z.$$

Equating indices, we obtain $x = -1$, $z = 1$, $y = 4$, that is,

$$\text{volume per second} = k.\frac{pa^4}{\eta l}.$$

Poiseuille, as shown below, found $k = \pi/8$, i.e.

$$\text{volume per second} = \pi p a^4/8\eta l.$$

Proof by calculus. For laminar flow, the resultant force on a cylinder of liquid is zero, i.e. $\eta A dv/dr + \pi r^2 p = 0$, where $A = 2\pi r l$. Integrating, and using $v = 0$ when $r = a$, we obtain

$$v = \frac{p}{4\eta l}(a^2 - r^2).$$

Considering a cylindrical shell of liquid, moving with velocity v, total volume per second through pipe $= \int_o^a v.2\pi r.dr = \frac{\pi p a^4}{8\eta l}$.

Turbulence. As the pressure difference between the ends of the pipe is increased, the liquid motion changes suddenly from an orderly to a turbulent motion. The velocity when this occurs is called the *critical velocity*. See Example 2, p. 163.

Reynolds demonstrated the change from orderly to turbulent motion by placing a wide pipe T in the bottom of a tank filled with water. (Fig. 17.2). The velocity of flow along T is controlled by a clip-on rubber tubing at the open end at C. A drawn-out glass jet B, connected to a reservoir A containing coloured water, is placed near the other end of T, and a thin jet of coloured water passes along the axis of T at low velocities. When the velocity reaches the critical value the colour spreads out and fills T, showing turbulence has begun (see Fig. 17.2*b*).

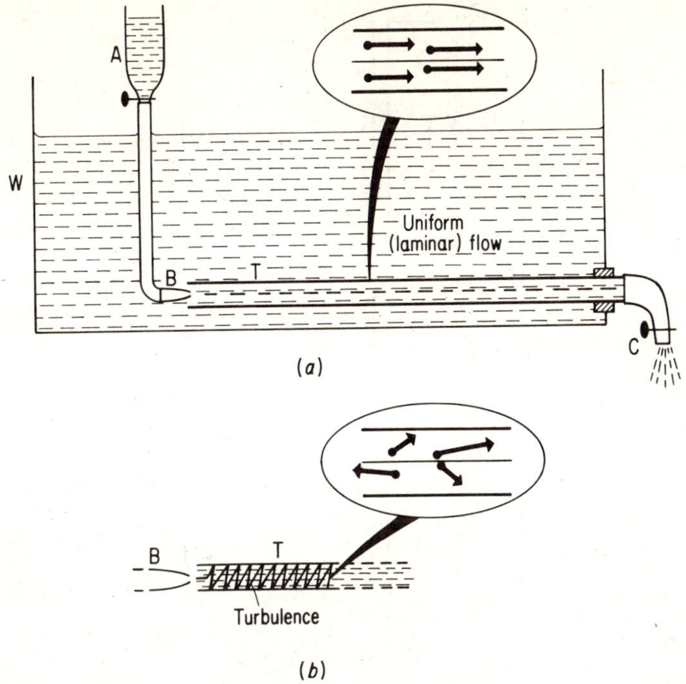

FIG. 17.2. Laminar (uniform) and turbulent flow.

Comparison of viscosities. The viscosity of oils and other liquids of low viscosity such as water can be compared by the *Ostwald viscometer*. This has a vertical capillary tube T, with two bulbs above T, and another wider tube from T is bent round to make a U-shaped apparatus (Fig. 17.3). *Equal volumes* of liquids are placed in turn in the viscometer by means of a pipette, drawn up to a suitable height, and the respective times t_1, t_2 to fall between the graduations P, Q of one of the bulbs are noted. Then, using Poiseuille's formula, it can be shown that

$$\frac{\eta_1}{\eta_2} = \frac{t_1}{t_2} \cdot \frac{\rho_1}{\rho_2},$$

where ρ_1, ρ_2 are the respective liquid densities. The formula is not true if equal volumes of the two liquids are not used.

Variation of viscosity with temperature. This can be found with an Ostwald viscometer by immersing the latter in water in a large cylinder, and varying the temperature of the water (Fig. 17.3). The viscosity of a liquid *decreases* as the temperature increases.

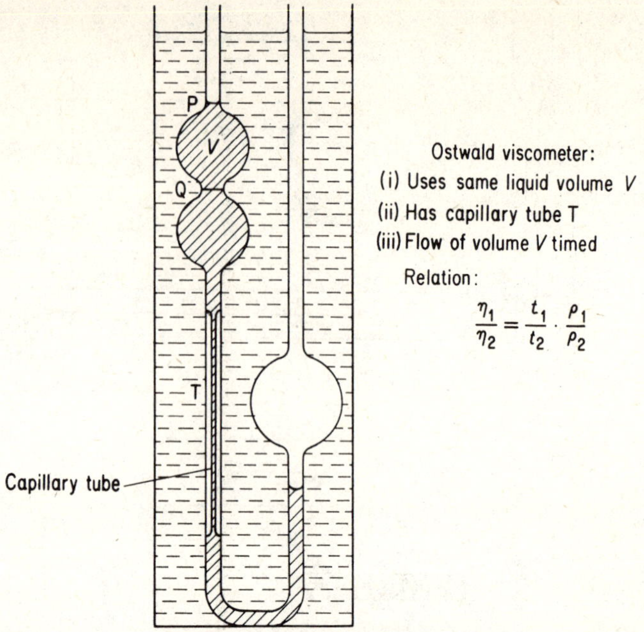

FIG. 17.3. Comparison of viscosities.

Absolute determination of viscosity of water. This is done by using a horizontal capillary tube T, along which water flows slowly under a small constant pressure head h of water (Fig. 17.4). The volume

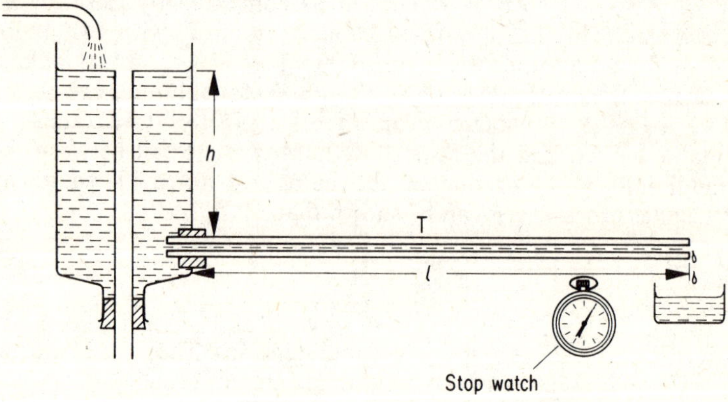

Poiseuille: vol. per second $= \frac{\pi(h\rho g)a^4}{8\eta l}$

FIG. 17.4. Absolute measurement of viscosity.

per second issuing from the tube is measured, and η calculated from the relation

$$\text{volume per second} = \frac{\pi p a^4}{8\eta l},$$

where $p = h\rho g$, a is the radius of the tube, and l is its length.

Viscosity of viscous liquids. The viscosity of glycerine or other very viscous liquids can be found by using *Stokes' law*. This states that the frictional force F on a sphere of radius a moving with velocity v through a fluid of viscosity η is given by $F = 6\pi a\eta v$.

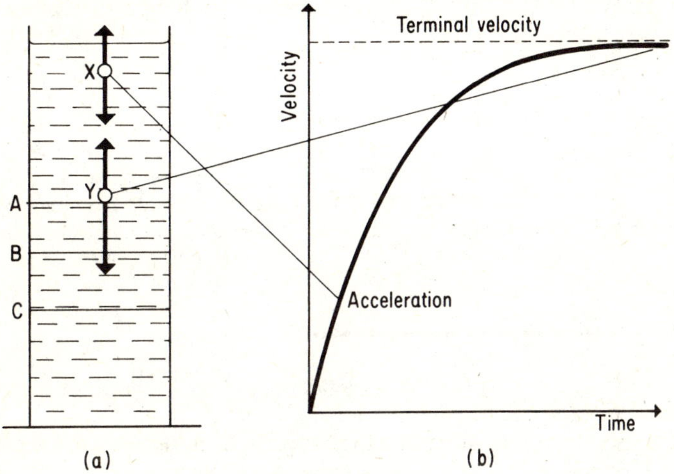

FIG. 17.5. Viscous force on sphere. Terminal velocity.

A sphere falling through a liquid accelerates at first (X), but soon reaches a terminal or constant velocity v_0 (Y) (Fig. 17.5*a*,*b*). The weight is then equal to the upthrust + frictional force, which acts upwards.

$$\therefore \tfrac{4}{3}\pi a^3 \rho g = 6\pi a\eta v_0 + \tfrac{4}{3}\pi a^3 \sigma g$$

$$\therefore v_0 = \frac{2ga^2(\rho-\sigma)}{9\eta}.$$

Thus η can be found if v_0, a, ρ, σ (liquid density) are known. The terminal velocity can be found by filling a tall gas jar with the liquid, dropping a small ball-bearing gently along the axis of the jar, and measuring the velocity by timing its fall over successive equal distances, AB, BC, in the middle of the jar, so as to check for terminal velocity.

Viscosity of gas. When a gas flows steadily along a pipe, fast and slower molecules constantly exchange layers, say between Y and X. (Fig. 17.6). The molecules carry their own momentum into a layer. Thus the slower layers tend to be speeded up and the faster layers to be slowed down. This is the origin of the frictional force in the gas. The viscosity is thus due to the continual exchange of momentum between layers.

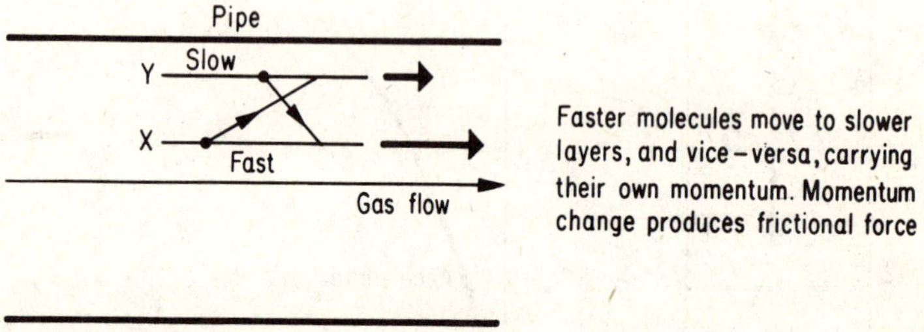

FIG. 17.6. Viscosity of gas.

Temperature rise increases the number of molecules per second crossing between layers. Hence the rate of momentum change increases. The viscosity of a gas thus rises with temperature rise. A rise in thermal energy of a *liquid*, however, enables the molecules in neighbouring layers to slide more easily over each other. Thus the viscosity of a liquid decreases with temperature rise.